TEACHER'S PLANNING GUIDE

Project-Based Inquiry Science™

DIGGING IN

84 Business Park Drive, Armonk, NY 10504
Phone (914) 273-2233 Fax (914) 273-2227
www.its-about-time.com

Program Components

Student Edition	**Durable Equipment Kit**
Teacher's Planning Guide	**Consumable Equipment Kit**
Teacher's Resources Guide	**Multimedia**
	— IDEO Deep Dive Video or IDEO Deep Dive DVD

All student activities in this textbook have been designed to be as safe as possible, and have been reviewed by professionals specifically for that purpose. As well, appropriate warnings concerning potential safety hazards are included where applicable to particular activities. However, responsibility for safety remains with the student, the classroom teacher, the school principal, and the school board.

Printed and bound in the United States of America.

ISBN 978-1-58591-628-3
1 2 3 4 5 CRS 12 11 10 09 08

This project was supported, in part, by the ***National Science Foundation*** under grant nos. 0137807, 0527341, and 0639978.
Opinions expressed are those of the authors and not necessarily those of the National Science Foundation.

Principal Investigators

Janet L. Kolodner is a Regents' Professor in the School of Interactive Computing in the Georgia Institute of Technology's College of Computing. Since 1978, her research has focused on learning from experience, both in computers and in people. She pioneered the Artificial Intelligence method called *case-based reasoning*, providing a way for computers to solve new problems based on their past experiences. Her book, *Case-Based Reasoning*, synthesizes work across the case-based reasoning research community from its inception to 1993.

Since 1994, Dr. Kolodner has focused on the applications and implications of case-based reasoning for education. In her approach to science education, called Learning by Design™ (LBD), students learn science while pursuing design challenges. Dr. Kolodner has investigated how to create a culture of collaboration and rigorous science talk in classrooms, how to use a project challenge to promote focus on science content, and how students learn and develop when classrooms function as learning communities. Currently, Dr. Kolodner is investigating how to help young people come to think of themselves as scientific reasoners. Dr. Kolodner's research results have been widely published, including in *Cognitive Science, Design Studies,* and the *Journal of the Learning Sciences.*

Dr. Kolodner was founding Director of Georgia Tech's EduTech Institute, served as coordinator of Georgia Tech's Cognitive Science program for many years, and is founding Editor in Chief of the *Journal of the Learning Sciences*. She is a founder of the International Society for the Learning Sciences, and she served as its first Executive Officer. She is a fellow of the American Association of Artificial Intelligence.

Joseph S. Krajcik is a Professor of Science Education and Associate Dean for Research in the School of Education at the University of Michigan. He works with teachers in science classrooms to bring about sustained change by creating classroom environments in which students find solutions to important intellectual questions that subsume essential curriculum standards and use learning technologies as productivity tools. He seeks to discover what students learn in such environments, as well as to explore and find solutions to challenges that teachers face in enacting such complex instruction.

Dr. Krajcik has authored and co-authored over 100 manuscripts and makes frequent presentations at international, national, and regional conferences that focus on his research, as well as presentations that translate research findings into classroom practice. He is a fellow of the American Association for the Advancement of Science and served as president of the National Association for Research in Science Teaching. Dr. Krajcik co-directs the Center for Highly Interactive Classrooms, Curriculum and Computing in Education at the University of Michigan and is a co-principal investigator in the Center for Curriculum Materials in Science and The National Center for Learning and Teaching Nanoscale Science and Engineering. In 2002, Dr. Krajcik was honored to receive a Guest Professorship from Beijing Normal University in Beijing, China. In winter 2005, he was the Weston Visiting Professor of Science Education at the Weizmann Institute of Science in Rehovot, Israel.

Daniel C. Edelson is Vice President for Education and Children's Programs at the National Geographic Society. Previously, he was the director of the Geographic Data in Education (GEODE) Initiative at Northwestern University, where he led the development of Planetary Forecaster and Earth Systems and Processes. Since 1992, Dr. Edelson has directed a series of projects exploring the use of technology as a catalyst for reform in science education and has led the development of a number of software environments for education. These include My World GIS, a geographic information system for inquiry-based learning, and WorldWatcher, a data visualization and analysis system for gridded geographic data. Dr. Edelson is the author of the high school environmental science text, *Investigations in Environmental Science: A Case-Based Approach to the Study of Environmental Systems*. His research has been widely published, including in the *Journal of the Learning Sciences*, the *Journal of Research on Science Teaching*, *Science Educator*, and *Science Teacher*.

Brian J. Reiser is a Professor of Learning Sciences in the School of Education and Social Policy at Northwestern University. Professor Reiser served as chair of Northwestern's Learning Sciences Ph.D. program from 1993, shortly after its inception, until 2001. His research focuses on the design and enactment of learning environments that support students' inquiry in science, including both science curriculum materials and scaffolded software tools. His research investigates the design of learning environments that scaffold scientific practices, including investigation, argumentation, and explanation; design principles for technology-infused curricula that engage students in inquiry projects; and the teaching practices that support student inquiry. Professor Reiser also directed BGuILE (Biology Guided Inquiry Learning Environments) to develop software tools for supporting middle school and high school students in analyzing data and constructing explanations with biological data. Reiser is a co-principal investigator in the NSF Center for Curriculum Materials in Science. He served as a member of the NRC panel authoring the report Taking Science to School.

Mary L. Starr is a Research Specialist in Science Education in the School of Education at the University of Michigan. She collaborates with teachers and students in elementary and middle school science classrooms around the United States who are implementing *Project-Based Inquiry Science*. Before joining the PBIS team, Dr. Starr created professional learning experiences in science, math, and technology, designed to assist teachers in successfully changing their classroom practices to promote student learning from coherent inquiry experiences. She has developed instructional materials in several STEM areas, including nanoscale science education, has presented at national and regional teacher education and educational research meetings, and has served in a leadership role in the Michigan Science Education Leadership Association. Dr. Starr has authored articles and book chapters, and has worked to improve elementary science teacher preparation through teaching science courses for pre-service teachers and acting as a consultant in elementary science teacher preparation. As part of the PBIS team, Dr. Starr has played a lead role in making units cohere as a curriculum, in developing the framework for PBIS Teacher's Planning Guides, and in developing teacher professional development experiences and materials.

Acknowledgements

Three research teams contributed to the development of *Project-Based Inquiry Science* (PBIS): a team at the Georgia Institute of Technology headed by Janet L. Kolodner, a team at Northwestern University headed by Daniel Edelson and Brian Reiser, and a team at the University of Michigan headed by Joseph Krajcik and Ron Marx. Each of the PBIS units was originally developed by one of these teams and then later revised and edited to be a part of the full three-year middle-school curriculum that became PBIS.

PBIS has its roots in two educational approaches, Project-Based Science and Learning by Design™. Project-Based Science suggests that students should learn science through engaging in the same kinds of inquiry practices scientists use, in the context of scientific problems relevant to their lives and using tools authentic to science. Project-Based Science was originally conceived in the hi-ce Center at the University of Michigan, with funding from the National Science Foundation. Learning by Design™ derives from Problem-Based Learning and suggests sequencing, social practices, and reflective activities for promoting learning. It engages students in design practices, including the use of iteration and deliberate reflection. LBD was conceived at the Georgia Institute of Technology, with funding from the National Science Foundation, DARPA, and the McDonnell Foundation.

The development of the integrated PBIS curriculum was supported by the National Science Foundation under grants no. 0137807, 0527341, and 0639978. Any opinions, findings and conclusions, or recommendations expressed in this material are those of the authors and do not necessarily reflect the views of the National Science Foundation.

PBIS Team

Principal Investigator
Janet L. Kolodner

Co-Principal Investigators
Daniel C. Edelson
Joseph S. Krajcik
Brian J. Reiser

NSF Program Officer
Gerhard Salinger

Curriculum Developers
Michael T. Ryan
Mary L. Starr

Teacher's Planning Guide Developers
Rebecca M. Schneider
Mary L. Starr

Literacy Specialist
LeeAnn M. Sutherland

NSF Program Reviewer
Arthur Eisenkraft

Project Coordinator
Juliana Lancaster

External Evaluators
The Learning Partnership
Steven M. McGee
Jennifer Witers

The Georgia Institute of Technology Team

Project Director:
Janet L. Kolodner

Development of PBIS units at the Georgia Institute of Technology was conducted in conjunction with the Learning by Design™ Research group (LBD), Janet L. Kolodner, PI.

Lead Developers, Physical Science:
David Crismond
Michael T. Ryan

Lead Developer, Earth Science:
Paul J. Camp

Assessment and Evaluation:
Barbara Fasse
Daniel Hickey
Jackie Gray
Laura Vandewiele
Jennifer Holbrook

Project Pioneers:
JoAnne Collins
David Crismond
Joanna Fox
Alice Gertzman
Mark Guzdial
Cindy Hmelo-Silver
Douglas Holton
Roland Hubscher
N. Hari Narayanan
Wendy Newstetter
Valery Petrushin
Kathy Politis
Sadhana Puntambekar
David Rector
Janice Young

The Northwestern University Team

Project Directors:
Daniel Edelson
Brian Reiser

Lead Developer, Biology:
David Kanter

Lead Developers, Earth Science:
Jennifer Mundt Leimberer
Darlene Slusher

Development of PBIS units at Northwestern was conducted in conjunction with:

The Center for Learning Technologies in Urban Schools (LeTUS) at Northwestern, and the Chicago Public Schools
Louis Gomez, PI;
Clifton Burgess, PI
for Chicago Public Schools.

The BioQ Collaborative
David Kanter, PI.

The Biology Guided Inquiry Learning Environments (BGuILE) Project
Brian Reiser, PI.

The Geographic Data in Education (GEODE) Initiative
Daniel Edelson, Director

The Center for Curriculum Materials in Science at Northwestern
Brian Reiser,
Daniel Edelson,
Bruce Sherin, PIs.

The University of Michigan Team

Project Directors:
Joseph Krajcik
Ron Marx

Literacy Specialist:
LeeAnn M. Sutherland

Project Coordinator:
Mary L. Starr

Development of PBIS units at the University of Michigan was conducted in conjunction with:

The Center for Learning Technologies in Urban Schools (LeTUS)
Ron Marx, Phyllis Blumenfeld,
Barry Fishman,
Joseph Krajcik,
Elliot Soloway, PIs.

The Detroit Public Schools
Juanita Clay-Chambers
Deborah Peek-Brown

The Center for Highly Interactive Computing in Education (hi-ce)
Ron Marx,
Phyllis Blumenfeld,
Barry Fishman,
Joseph Krajcik,
Elliot Soloway,
Elizabeth Moje,
LeeAnn Sutherland, PIs.

Field-Test Teachers

National Field Test

Tamica Andrew
Leslie Baker
Jeanne Bayer
Gretchen Bryant
Boris Consuegra
Daun D'Aversa
Candi DiMauro
Kristie L. Divinski
Donna M. Dowd
Jason Fiorito
Lara Fish
Christine Gleason
Christine Hallerman
Terri L. Hart-Parker
Jennifer Hunn
Rhonda K. Hunter
Jessica Jones
Dawn Kuppersmith
Anthony F. Lawrence
Ann Novak
Rise Orsini
Tracy E. Parham
Cheryl Sgro-Ellis
Debra Tenenbaum
Sarah B. Topper
Becky Watts
Debra A. Williams
Ingrid M. Woolfolk
Ping-Jade Yang

New York City Field Test

Several sequences of PBIS units have been field- tested in New York City under the leadership of Whitney Lukens, Staff Developer for Region 9, and Greg Borman, Science Instructional Specialist, New York City Department of Education

6th Grade

Norman Agard
Tazinmudin Ali
Heather Guthartz Aniba
Asher Arzonane
Asli Aydin
Shareese Blakely
John J. Blaylock
Joshua Blum
Tsedey Bogale
Filomena Borrero
Zachary Brachio
Thelma Brown
Alicia Browne-Jones
Scott Bullis
Maximo Cabral
Lionel Callender
Matthew Carpenter
Ana Maria Castro
Diane Castro
Anne Chan
Ligia Chiorean
Boris Consuegra
Careen Halton Cooper
Cinnamon Czarnecki
Kristin Decker
Nancy Dejean
Gina DiCicco
Donna Dowd
Lizanne Espina
Joan Ferrato
Matt Finnerty
Jacqueline Flicker
Helen Fludd
Leigh Summers Frey
Helene Friedman-Hager
Diana Gering
Matthew Giles
Lucy Gill
Steven Gladden
Greg Grambo
Carrie Grodin-Vehling
Stephan Joanides
Kathryn Kadei
Paraskevi Karangunis
Cynthia Kerns
Martine Lalanne
Erin Lalor
Jennifer Lerman
Sara Lugert
Whitney Lukens
Dana Martorella
Christine Mazurek
Janine McGeown
Chevelle McKeever
Kevin Meyer
Jennifer Miller
Nicholas Miller
Diana Neligan
Caitlin Van Ness
Marlyn Orque
Eloisa Gelo Ortiz
Gina Papadopoulos
Tim Perez
Albertha Petrochilos
Christopher Poli
Kristina Rodriguez
Nadiesta Sanchez
Annette Schavez
Hilary Sedgwitch
Elissa Seto
Laura Shectman
Audrey Shmuel
Katherine Silva
Ragini Singhal
C. Nicole Smith
Gitangali Sohit
Justin Stein
Thomas Tapia
Eilish Walsh-Lennon
Lisa Wong
Brian Yanek
Cesar Yarleque
David Zaretsky
Colleen Zarinsky

7th Grade

Mayra Amaro
Emmanuel Anastasiou
Cheryl Barnhill
Bryce Cahn
Ligia Chiorean
Ben Colella
Boris Consuegra
Careen Halton Cooper
Elizabeth Derse
Urmilla Dhanraj
Gina DiCicco
Lydia Doubleday
Lizanne Espina
Matt Finnerty
Steven Gladden
Stephanie Goldberg
Nicholas Graham
Robert Hunter
Charlene Joseph
Ketlynne Joseph
Kimberly Kavazanjian
Christine Kennedy
Bakwah Kotung
Lisa Kraker
Anthony Lett
Herb Lippe
Jennifer Lopez
Jill Mastromarino
Kerry McKie
Christie Morgado
Patrick O'Connor
Agnes Ochiagha
Tim Perez
Nadia Piltser
Chris Poli
Carmelo Ruiz
Kim Sanders
Leslie Schiavone
Ileana Solla
Jacqueline Taylor
Purvi Vora
Ester Wiltz
Carla Yuille
Marcy Sexauer Zacchea
Lidan Zhou

8th Grade

Emmanuel Anastasio
Jennifer Applebaum
Marsha Armstrong
Jenine Barunas
Vito Cipolla
Kathy Critharis
Patrecia Davis
Alison Earle
Lizanne Espina
Matt Finnerty
Ursula Fokine
Kirsis Genao
Steven Gladden
Stephanie Goldberg
Peter Gooding
Matthew Herschfeld
Mike Horowitz
Charlene Jenkins
Ruben Jimenez
Ketlynne Joseph
Kimberly Kavazanjian
Lisa Kraker
Dora Kravitz
Anthony Lett
Emilie Lubis
George McCarthy
David Mckinney
Michael McMahon
Paul Melhado
Jen Miller
Christie Morgado
Ms. Oporto
Maria Jenny Pineda
Anastasia Plaunova
Carmelo Ruiz
Riza Sanchez
Kim Sanders
Maureen Stefanides
Dave Thompson
Matthew Ulmann
Maria Verosa
Tony Yaskulski

DIGGING IN

Digging In was originally developed at the Georgia Institute of Technology as part of the Learning by Design™ initiative and was titled *Earth Science: Digging In.*

Digging In

PBIS Editorial Team:
Michael T. Ryan
Victoria Deneroff

Georgia Institute of Technology Team

Project Director:
Janet L. Kolodner

Lead Authors:
Paul Camp
Jennifer Holbrook

Contributing Authors:
David Crismond
Mike Ryan
Jennifer Turns

Formative Development:
David Crismond
Joanna Fox
Jackie Gray
Cami Heck
Jennifer Holbrook
Susan McClendon
Kristine Nagel
Lindy Wine
Janice Young

Pilot Teachers:
Barbara Blasch
Audrey Daniel
Emily Dickson
Carmen Dillard
Yvette Fernandez
Joyce Gamble
Dorothy Hicks
Daphne Islam-Gordon
Rudo Kashiri
Marni Klein
Toni Laman
Paige Lefont
Susan McClendon
Bernie Moore
Kelly Rowsey
Mike Ryan
Beth Smith
Delilah Springer
Lindy Wine
Avis Winfield
Mary Winn
Ann Yergin

The development of *Earth Science: Digging In* was supported by the National Science Foundation under grant nos. 9553583, 9818828 and 0208059 and by grants from the McDonell Foundation, the BellSouth Foundation, the Woodruff Foundation, and the Georgia Tech Foundation. Any opinions, findings, and conclusions or recommendations expressed in this material are those of the authors and do not necessarily reflect the views of the National Science Foundation.

Digging In Teacher's Planning Guide

Learning Set 2

Science Concepts: *Models, simulations, designing and measuring procedures, standardized procedures, range of results, inconsistent data, reliability, variation, line plots, data, trials, precision, volcanoes, magma, lava, types of lava, rock, rock classification, minerals.*

Learning Set 3

Science Concepts: *Criteria and constraints, recording observations, phenomena, repeatable, replicating results, interpretation, trends, claims, fair test, variables, independent variable, control variables, dependent variables, evidence, ingenuity, explanations, cases and case studies, erosion case studies, erosion, deposition, factors that affect erosion.*

3.0 Learning Set Introduction

The Basketball-Court Challenge

3.1 Understand the Question

Thinking About Erosion

3.2 Case Studies

What Causes Erosion?

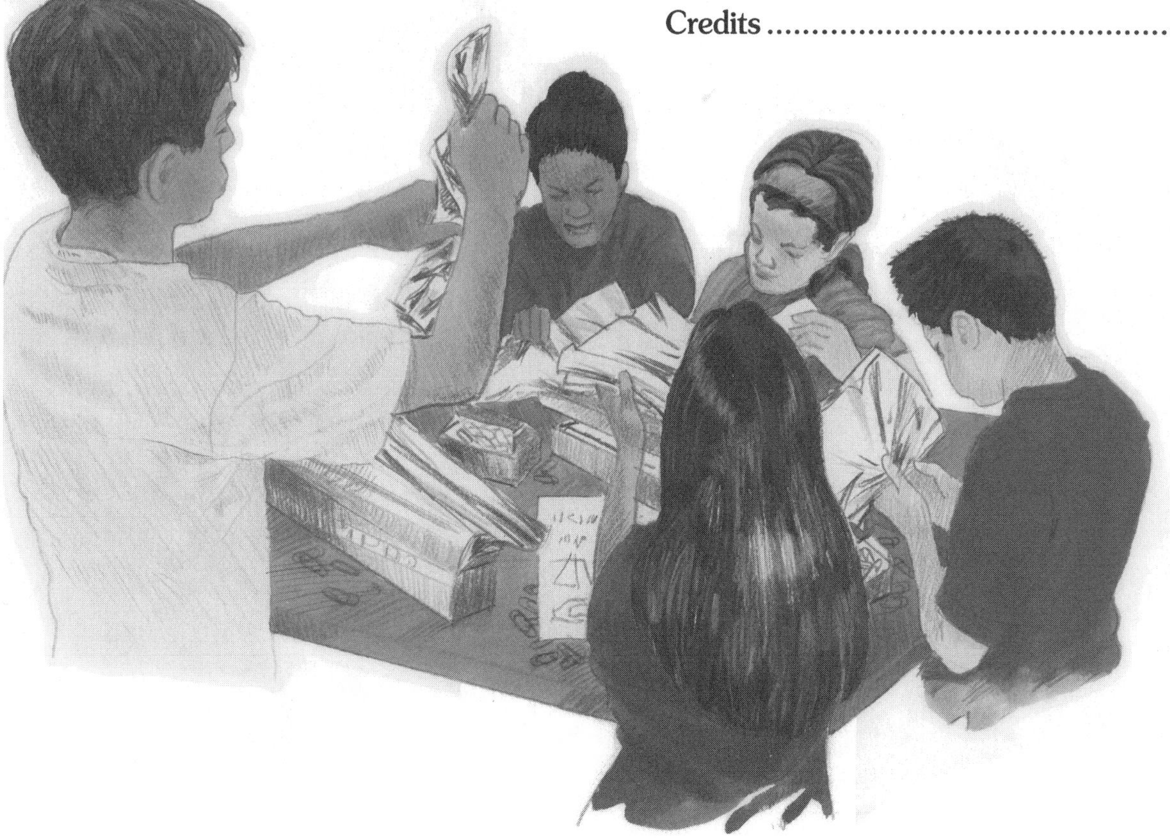

Welcome to Project-Based Inquiry Science!

Welcome to Project-Based Inquiry Science (PBIS): A Middle-School Science Curriculum!

This year, your students will be learning the way scientists learn, exploring interesting questions and challenges, reading about what other scientists have discovered, investigating, experimenting, gathering evidence, and forming explanations. They will learn to collaborate with others to find answers and to share their learning in a variety of ways. In the process, they will come to see science in a whole new, exciting way that will motivate them throughout their educational experiences and beyond.

What is PBIS?

In project-based inquiry learning, students investigate scientific content and learn science practices in the context of attempting to address challenges in or answer questions about the world around them. Early activities introducing students to a challenge help them to generate issues that need to be investigated, making inquiry a student-driven endeavor. Students investigate as scientists would, through observations, designing and running experiments, designing, building, and running models, reading written material, and so on, as appropriate. Throughout each project, students might make use of technology and computer tools that support their efforts in observation, experimentation, modeling, analysis, and reflection. Teachers support and guide the student inquiries by framing the guiding challenge or question, presenting crucial lessons, managing the sequencing of activities, and eliciting and steering discussion and collaboration among the students. At the completion of a project, students publicly exhibit what they have learned along with their solutions to the specific challenge. Personal reflection to help students learn from the experience is embedded in student activities, as are opportunities for assessment.

The curriculum will provide three years of piloted project-based inquiry materials for middle-school science. Individual curriculum units have been defined that cover the scope of the national content and process standards for the middle-school grades. Each Unit focuses on helping students acquire qualitative understanding of targeted science principles and move toward quantitative understanding, is infused with technology, and provides a foundation in reasoning skills, science content, and science process that will ready them for more advanced science. The curriculum as a whole introduces students to a wide range of investigative approaches in science (e.g., experimentation, modeling) and is designed to help them develop scientific reasoning skills that span those investigative approaches.

Technology can be used in project-based inquiry to make available to students some of the same kinds of tools and aids used by scientists in the field. These range from pencil-and-paper tools for organized data recording, collection, and management to software tools for analysis, simulation, modeling, and other tasks. Such infusion provides a platform for providing prompts, hints, examples, and other kinds of aids to students as they are engaging in scientific reasoning. The learning technologies and tools that are integrated into the curriculum offer essential scaffolding to students as they are developing their scientific reasoning skills, and are seamlessly infused into the overall completion of project activities and investigations.

Standards-Based Development

Development of each curriculum Unit begins by identifying the specific relevant national standards to be addressed. Each Unit has been designed to cover a specific portion of the national standards. This phase of development also includes an analysis of curriculum requirements across multiple states. Our intent is to deliver a product that will provide coverage of the content deemed essential on the widest practical scope and that will be easily adaptable to the needs of teachers across the country.

Once the appropriate standards have been identified, the development team works to define specific learning goals built from those standards, and takes into account conceptions and misunderstandings common among middle-school students. An orienting design challenge or driving question for investigation is chosen that motivates achieving those learning goals, and the team then sequences activities and the presentation of specific concepts so that students can construct an accurate understanding of the subject matter.

Inquiry-Based Design

The individual curriculum Units present two types of projects: engineering-design challenges and driving-question investigations. Design-challenge Units begin by presenting students with a scenario and problem and challenging them to design a device or plan that will solve the problem. Driving-question investigations begin by presenting students with a complex question with real-world implications. Students are challenged to develop answers to the questions. The scenario and problem in the design Units and the driving question in the investigation Units are carefully selected to lead the students into investigation of specific science concepts, and the solution processes are carefully structured to require use of specific scientific reasoning skills.

Pedagogical Rationale

Research shows that individual project-based learning units promote excitement and deep learning of the targeted concepts. However, achieving deep, flexible, transferable learning of cross-disciplinary content (e.g., the notion of a model, time scale, variable, experiment) and science practice requires a learning environment that consistently, persistently, and pervasively encourages the use of such content and practices over an extended period of time. By developing project-based inquiry materials that cover the spectrum of middle-school science content in a coherent framework, we provide this extended exposure to the type of learning environment most likely to produce competent scientific thinkers who are well grounded in their understanding of both basic science concepts and the standards and practices of science in general.

Evidence of Effectiveness

There is compelling evidence showing that a project-based inquiry approach meets this goal. Working at Georgia Tech, the University of Michigan, and Northwestern University, we have developed, piloted, and/or field-tested many individual project-based units. Our evaluation evidence shows that these materials engage students well and are manageable by teachers, and that students learn both content and process skills. In every summative evaluation, student performance on post-tests improved significantly from pretest performance (Krajcik, et al., 2000; Holbrook, et al., 2001; Gray et. al. 2001). For example, in the second year in a project-based classroom in Detroit, the average student at post-test scored at about the 95th percentile of the pre-test distribution. Further, we have repeatedly documented significant gains in content knowledge relative to other inquiry-based (but not project-based) instructional methods. In one set of results, performance by a project-based class

in Atlanta doubled on the content test while the matched comparison class (with an excellent teacher) experienced only a 20% gain (significance $p < .001$). Other comparisons have shown more modest differences, but project-based students consistently perform better than their comparisons. Most exciting about the Atlanta results is that results from performance assessments show that, within comparable student populations, project-based students score higher on all categories of problem-solving and analysis and are more sophisticated at science practice and managing a collaborative scientific investigation. Indeed, the performance of average-ability project-based students is often statistically indistinguishable from or better than performance of comparison honors students learning in an inquiry-oriented but not project-based classroom. The Chicago group also has documented significant change in process skills in project-based classrooms. Students become more effective in constructing and critiquing scientific arguments (Sandoval, 1998) and in constructing scientific explanations using discipline-specific knowledge, such as evolutionary explanations for animal behavior (Smith & Reiser, 1998).

Researchers at Northwestern have also investigated the changes in classroom practices that are elicited by project-based units. Analyses of the artifacts students produce indicate that students are engaging in ambitious learning practices, requiring weighing and synthesizing many results from complex analyses of data, and constructing scientific arguments that require synthesizing results from multiple complex analyses of data (Edelson et al, 1998; Reiser et al, 2001). Students are engaged in planning, performing, monitoring and revising their investigations, and reporting on their investigation processes as well as their results (Loh et al, 1998). In general, the classrooms engaging in project-based activities reveal substantial moves toward a scientific discourse community in which students focus on arguing from evidence, critiquing ideas, and conjecturing, rather than simply reporting on what they have read or been told (Tabak & Reiser, 1997).

Introducing PBIS

What Do Scientists Do?

1) Scientists...address big challenges and big questions.

Students will find many different kinds of *Big Challenges* and *Questions* in *PBIS* Units. Some ask them to think about why something is a certain way. Some ask them to think about what causes something to change. Some challenge them to design a solution to a problem. Most are about things that can and do happen in the real world.

Understand the Big Challenge or Question

As students get started with each Unit, they will do activities that help them understand the *Big Question* or *Challenge* for that Unit. They will think about what they already know that might help them, and they will identify some of the new things they will need to learn.

Project Board

The *Project Board* helps you and your students keep track of their learning. For each challenge or question, they will use a *Project Board* to keep track of what they know, what they need to learn, and what they are learning. As they learn and gather evidence, they will record that on the *Project Board*. After they have answered each small question or challenge, they will return to the *Project Board* to record how what they have learned helps them answer the *Big Question* or *Challenge*.

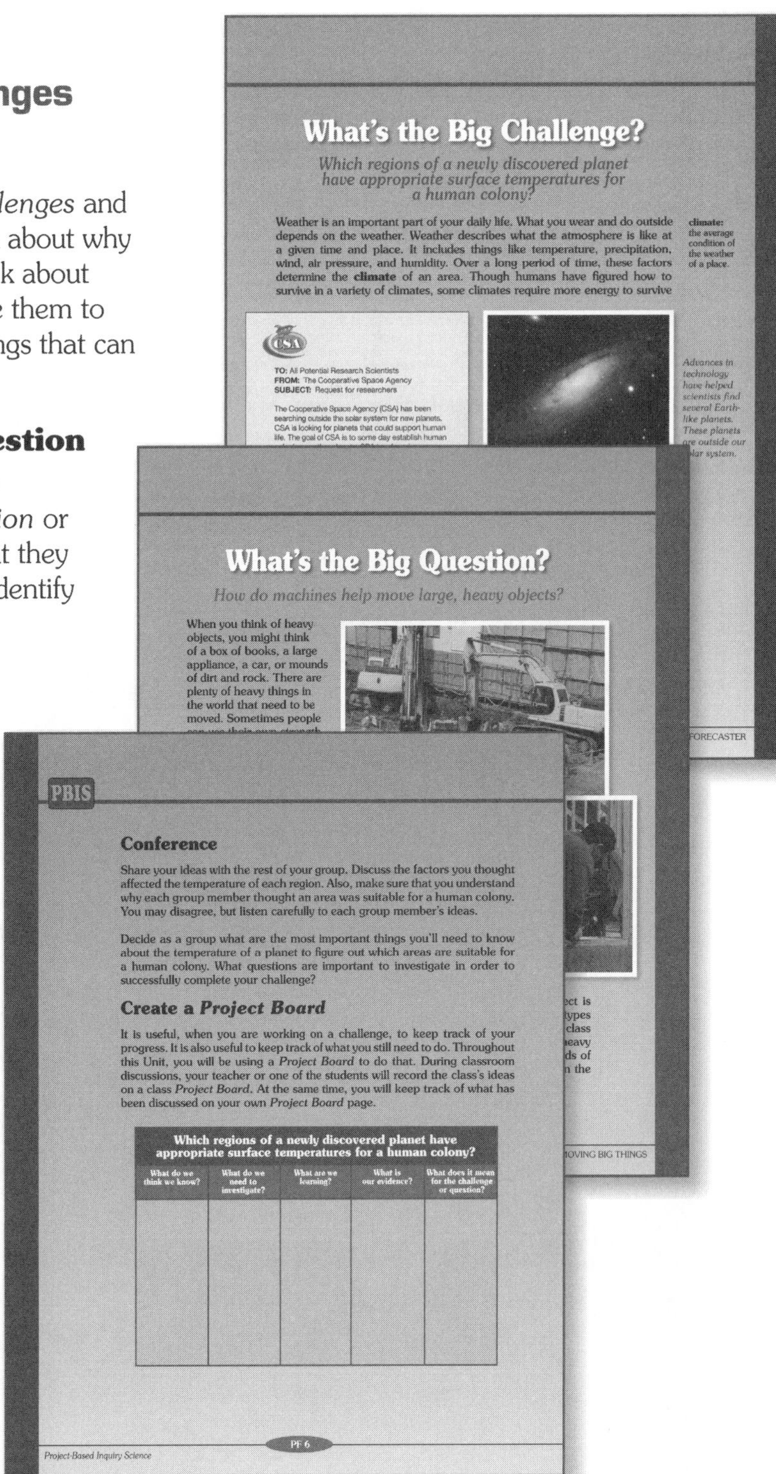
What's the Big Challenge?

Which regions of a newly discovered planet have appropriate surface temperatures for a human colony?

Weather is an important part of your daily life. What you wear and do outside depends on the weather. Weather describes what the atmosphere is like at a given time and place. It includes things like temperature, precipitation, wind, air pressure, and humidity. Over a long period of time, these factors determine the **climate** of an area. Though humans have figured how to survive in a variety of climates, some climates require more energy to survive

climate: the average condition of the weather of a place.

CSA

TO: All Potential Research Scientists
FROM: The Cooperative Space Agency
SUBJECT: Request for researchers

The Cooperative Space Agency (CSA) has been searching outside the solar system for new planets. CSA is looking for planets that could support human life. The goal of CSA is to some day establish human

Advances in technology have helped scientists find several Earth-like planets. These planets are outside our solar system.

What's the Big Question?

How do machines help move large, heavy objects?

When you think of heavy objects, you might think of a box of books, a large appliance, a car, or mounds of dirt and rock. There are plenty of heavy things in the world that need to be moved. Sometimes people

PBIS

Conference

Share your ideas with the rest of your group. Discuss the factors you thought affected the temperature of each region. Also, make sure that you understand why each group member thought an area was suitable for a human colony. You may disagree, but listen carefully to each group member's ideas.

Decide as a group what are the most important things you'll need to know about the temperature of a planet to figure out which areas are suitable for a human colony. What questions are important to investigate in order to successfully complete your challenge?

Create a *Project Board*

It is useful, when you are working on a challenge, to keep track of your progress. It is also useful to keep track of what you still need to do. Throughout this Unit, you will be using a *Project Board* to do that. During classroom discussions, your teacher or one of the students will record the class's ideas on a class *Project Board*. At the same time, you will keep track of what has been discussed on your own *Project Board* page.

Which regions of a newly discovered planet have appropriate surface temperatures for a human colony?				
What do we think we know?	What do we need to investigate?	What are we learning?	What is our evidence?	What does it mean for the challenge or question?

PF 6

Project-Based Inquiry Science

Learning Set 1

How Do Flowing Water and Land Interact in a Community?

The big question for this unit is *How does water quality affect the ecology of a community?* So far you have considered what you already know about what water quality is. Now you may be wondering where the water you use comes from. If you live in a city or town, the water you use may come from a river. You would want to know the quality of the water you are using. To do so, it is important to know how the water gets into the river. You also need to know what happens to the water as the river flows across the land.

You may have seen rivers or other water bodies near your home, your school, or in your city. Think about the river closest to where you live. Consider from where the water in the river comes. If you have traveled along the river, think about what the land around the river looks like. Try to figure out what human activities occur in the area. Speculate as to whether these activities affect the quality of water in the river.

To answer the big question, you need to break it down into smaller questions. In this *Learning Set*, you will investigate two smaller questions. As you will discover, these questions are very closely related and very hard to separate. The smaller questions are *How does water affect the land as it moves through the community?* and *How does land use affect water*

PBIS

Address the Big Challenge

How Do Scientists Work Together to Solve Problems?

You began this unit with the question, *how do scientists work together to solve problems?* You did several small challenges. As you worked on those challenges you learned about how scientists solve problems. You will now watch a video about real-life designers. You will see what the people in the video are doing that is like what you have been doing. Then you will think about all the different things you have been doing during this unit. Lastly, you will write about what you have learned about doing science and being a scientist.

Watch

IDEO Video

The video you will watch follows a group of designers at IDEO. IDEO is an innovation and design firm. In the video, they face the challenge of designing and building a new kind of shopping cart. These designers are doing many of the same things that you did. They also use other practices that you did not use. As you watch the video, record the interesting things you see.

After watching the video, answer the questions on the next page. You might want to look at them before you watch the video. Answering these questions should help you answer the big question of this unit: *How do scientists work together to solve problems?*

100

Learning Sets

Each Unit is composed of a group of *Learning Sets*, one for each of the smaller questions that needs to be answered to address the *Big Question* or *Challenge*. In each *Learning Set*, students will investigate and read to find answers to the *Learning Set's* question. They will also have a chance to share the results of their investigations with their classmates and work together to make sense of what they are learning. As students come to understand answers to the questions on the *Project Board*, you will record those answers and the evidence they collected. At the end of each *Learning Set*, they will apply their knowledge to the *Big Question* or *Challenge*.

Answer the Big Question/ Address the Big Challenge

At the end of each Unit, students will put everything they have learned together to tackle the *Big Question* or *Challenge*.

2) Scientists...address smaller questions and challenges.

What Students Do in a Learning Set

Understanding the Question or Challenge

At the start of each *Learning Set*, students will usually do activities that will help them understand the *Learning Set's* question or challenge and recognize what they already know that can help them answer the question or achieve the challenge. Usually, they will visit the *Project Board* after these activities and record on it the even smaller questions that they need to investigate to answer a *Learning Set's* question.

Investigate/Explore

There are many different kinds of investigations students might do to find answers to questions. In the *Learning Sets,* they might

- design and run experiments;
- design and run simulations;
- design and build models;
- examine large sets of data.

Don't worry if your students haven't done these things before. The text will provide them with lots of help in designing their investigations and in analyzing thier data.

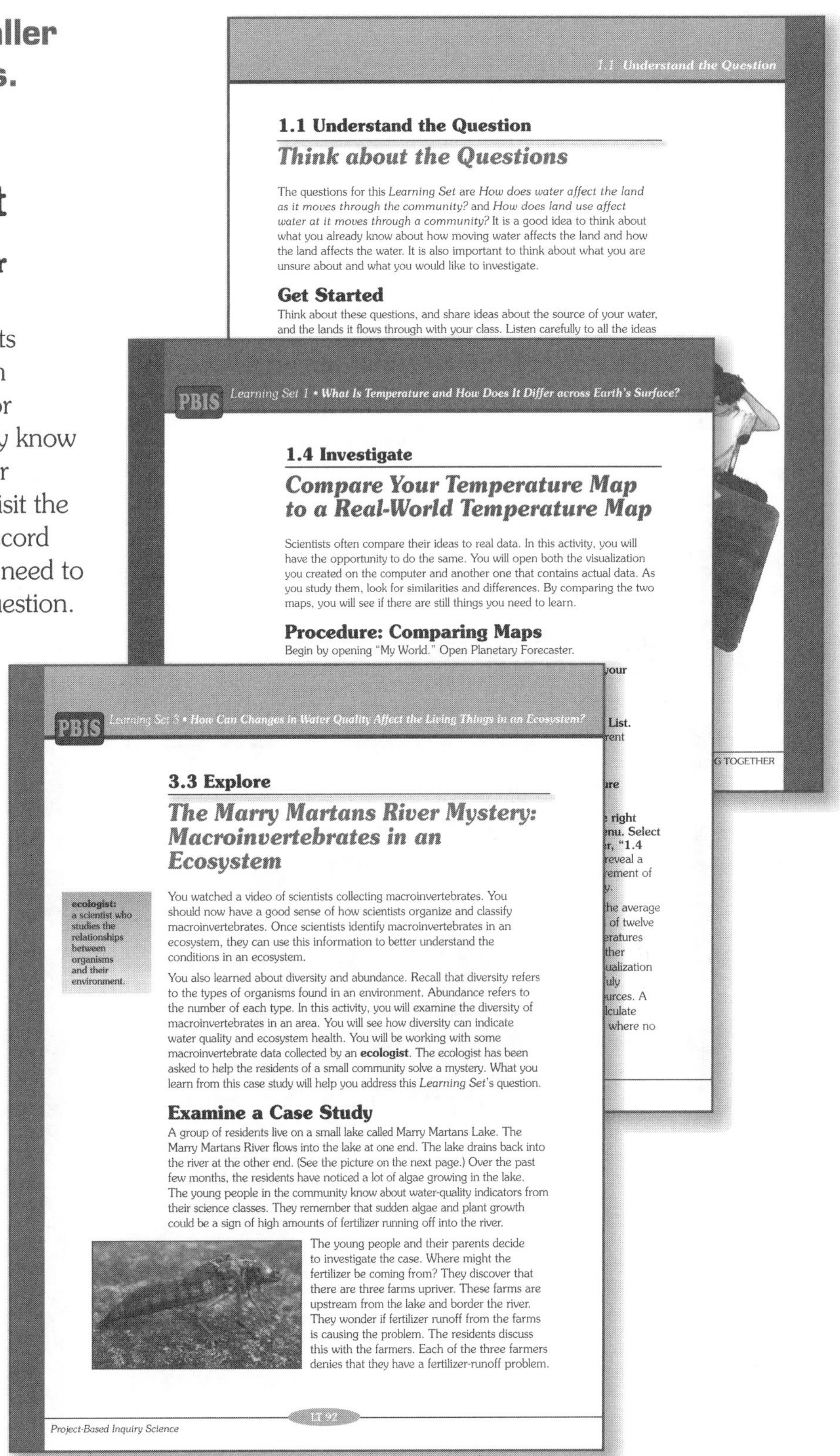

1.1 Understand the Question

1.1 Understand the Question

Think about the Questions

The questions for this *Learning Set* are *How does water affect the land as it moves through the community?* and *How does land use affect water at it moves through a community?* It is a good idea to think about what you already know about how moving water affects the land and how the land affects the water. It is also important to think about what you are unsure about and what you would like to investigate.

Get Started

Think about these questions, and share ideas about the source of your water, and the lands it flows through with your class. Listen carefully to all the ideas

PBIS *Learning Set 1* • *What Is Temperature and How Does It Differ across Earth's Surface?*

1.4 Investigate

Compare Your Temperature Map to a Real-World Temperature Map

Scientists often compare their ideas to real data. In this activity, you will have the opportunity to do the same. You will open both the visualization you created on the computer and another one that contains actual data. As you study them, look for similarities and differences. By comparing the two maps, you will see if there are still things you need to learn.

Procedure: Comparing Maps

Begin by opening "My World." Open Planetary Forecaster.

PBIS *Learning Set 3* • *How Can Changes in Water Quality Affect the Living Things in an Ecosystem?*

3.3 Explore

The Marry Martans River Mystery: Macroinvertebrates in an Ecosystem

ecologist: a scientist who studies the relationships between organisms and their environment.

You watched a video of scientists collecting macroinvertebrates. You should now have a good sense of how scientists organize and classify macroinvertebrates. Once scientists identify macroinvertebrates in an ecosystem, they can use this information to better understand the conditions in an ecosystem.

You also learned about diversity and abundance. Recall that diversity refers to the types of organisms found in an environment. Abundance refers to the number of each type. In this activity, you will examine the diversity of macroinvertebrates in an area. You will see how diversity can indicate water quality and ecosystem health. You will be working with some macroinvertebrate data collected by an **ecologist**. The ecologist has been asked to help the residents of a small community solve a mystery. What you learn from this case study will help you address this *Learning Set*'s question.

Examine a Case Study

A group of residents live on a small lake called Marry Martans Lake. The Marry Martans River flows into the lake at one end. The lake drains back into the river at the other end. (See the picture on the next page.) Over the past few months, the residents have noticed a lot of algae growing in the lake. The young people in the community know about water-quality indicators from their science classes. They remember that sudden algae and plant growth could be a sign of high amounts of fertilizer running off into the river.

The young people and their parents decide to investigate the case. Where might the fertilizer be coming from? They discover that there are three farms upriver. These farms are upstream from the lake and border the river. They wonder if fertilizer runoff from the farms is causing the problem. The residents discuss this with the farmers. Each of the three farmers denies that they have a fertilizer-runoff problem.

Project-Based Inquiry Science LT 92

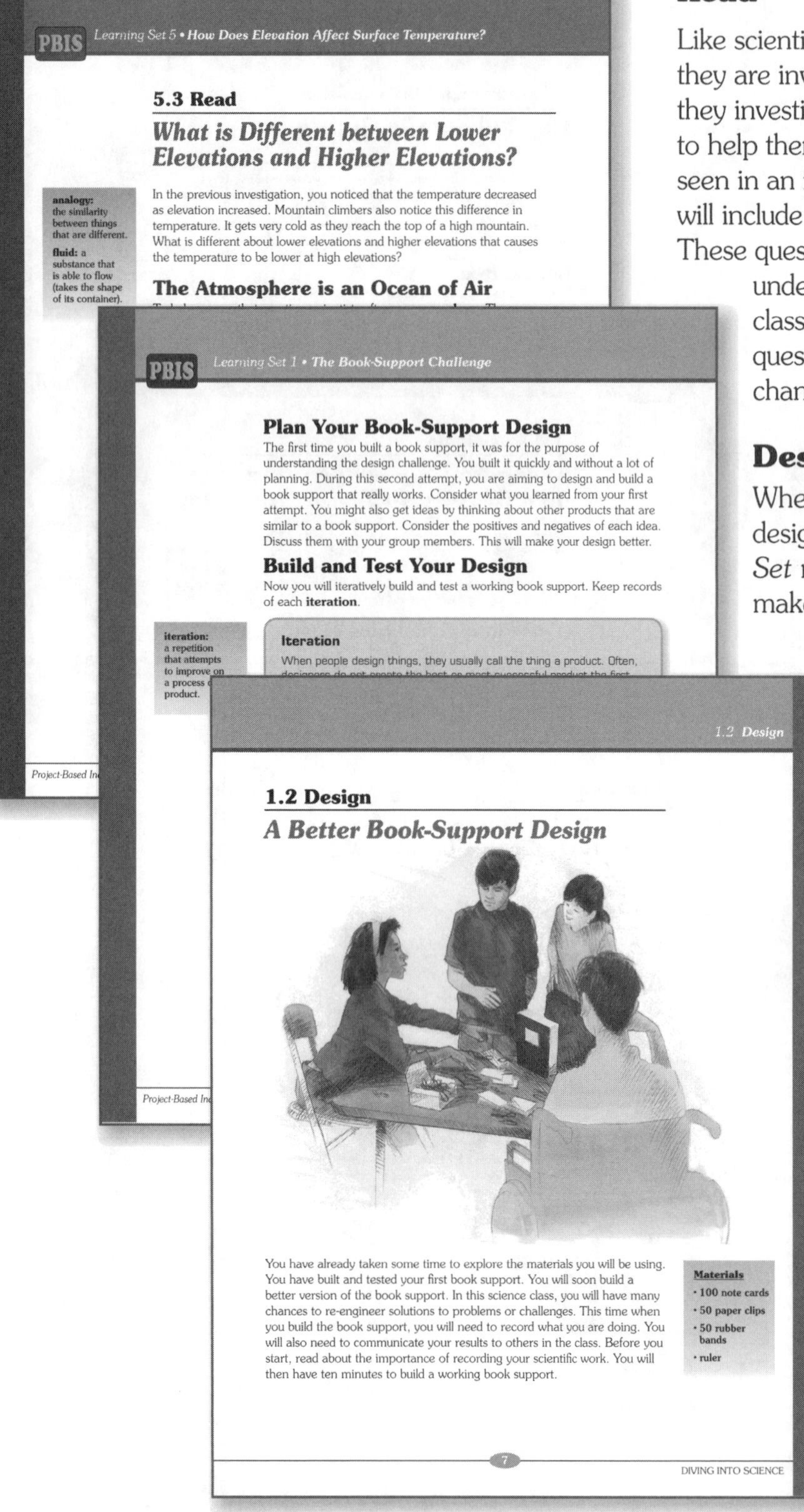

PBIS Learning Set 5 • How Does Elevation Affect Surface Temperature?

5.3 Read

What is Different between Lower Elevations and Higher Elevations?

analogy: the similarity between things that are different.

fluid: a substance that is able to flow (takes the shape of its container).

In the previous investigation, you noticed that the temperature decreased as elevation increased. Mountain climbers also notice this difference in temperature. It gets very cold as they reach the top of a high mountain. What is different about lower elevations and higher elevations that causes the temperature to be lower at high elevations?

The Atmosphere is an Ocean of Air

PBIS Learning Set 1 • The Book-Support Challenge

Plan Your Book-Support Design

The first time you built a book support, it was for the purpose of understanding the design challenge. You built it quickly and without a lot of planning. During this second attempt, you are aiming to design and build a book support that really works. Consider what you learned from your first attempt. You might also get ideas by thinking about other products that are similar to a book support. Consider the positives and negatives of each idea. Discuss them with your group members. This will make your design better.

Build and Test Your Design

Now you will iteratively build and test a working book support. Keep records of each **iteration**.

iteration: a repetition that attempts to improve on a process or product.

Iteration

When people design things, they usually call the thing a product. Often,

1.2 Design

1.2 Design

A Better Book-Support Design

You have already taken some time to explore the materials you will be using. You have built and tested your first book support. You will soon build a better version of the book support. In this science class, you will have many chances to re-engineer solutions to problems or challenges. This time when you build the book support, you will need to record what you are doing. You will also need to communicate your results to others in the class. Before you start, read about the importance of recording your scientific work. You will then have ten minutes to build a working book support.

Materials
- 100 note cards
- 50 paper clips
- 50 rubber bands
- ruler

7 DIVING INTO SCIENCE

Read

Like scientists, students will also read about the science they are investigating. They will read a little bit before they investigate, but most of the reading they do will be to help them understand what they have experienced or seen in an investigation. Each time they read, the text will include *Stop and Think* questions after the reading. These questions will help students gauge how well they understand what they have read. Usually, the class will discuss the answers to *Stop and Think* questions before going on so that everybody has a chance to make sense of the reading.

Design and Build

When the *Big Challenge* for a Unit asks them to design something, the challenge in a *Learning Set* might also ask them to design something and make it work. Often students will design a part of the thing they will design and build for the *Big Challenge*. When a *Learning Set* challenges students to design and build something, they will do several things:

- identify what questions they need to answer to be successful
- investigate to find answers to those questions
- use those answers to plan a good design solution
- build and test their design

Because designs don't always work the way one wants them to, students will usually do a design challenge more than once. Each time through, they will test their design. If their design doesn't work as well as they would like, they will determine why it is not working and identify other things they need to investigate to make it work better. Then, they will learn those things and try again.

Explain and Recommend

A big part of what scientists do is explain, or try to make sense of why things happen the way they do. An explanation describes why something is the way it is or behaves the way it does. An explanation is a statement one makes built from claims (what you think you know), evidence (from an investigation) that supports the claim, and science knowledge. As they learn, scientists get better at explaining. You will see that students get better, too, as they work through the *Learning Sets*.

A recommendation is a special kind of claim—one where you advise somebody about what to do. Students will make recommendations and support them with evidence, science knowledge, and explanations.

3.5 Explain

3.5 Explain

Create an Explanation

After scientists get results from an investigation, they try to make a claim. They base their claim on what their evidence shows. They also use what they already know to make their claim. They explain why their claim is valid. The purpose of a science explanation is to help others understand the following:

- what was learned from a set of investigations
- why the scientists reached this conclusion

Later, other scientists will use these explanations to help them explain other phenomena. The explanations will also help them predict what will happen in other situations.

You will do the same thing now. Your claim will be the trend you found in your experiment. You will use data you collected and science knowledge you have read to create a good explanation. This will help you decide whether your claim is valid. You will be reporting the results of the investigation to your classmates. With a good explanation that matches your claim, you can convince them that your claim is valid.

Because your understanding of the science of forces is not complete, you may not be able to fully explain your results. But you will use what you have read to come up with your best explanation. Scientists finding out about new things do the same thing. When they only partly understand something, it is impossible for them to form a "perfect" explanation. They do the best they can based on what they understand. As they learn more, they make their explanations better. This is what you will do now and what you will be doing throughout PBIS. You will explain your results the best you can based

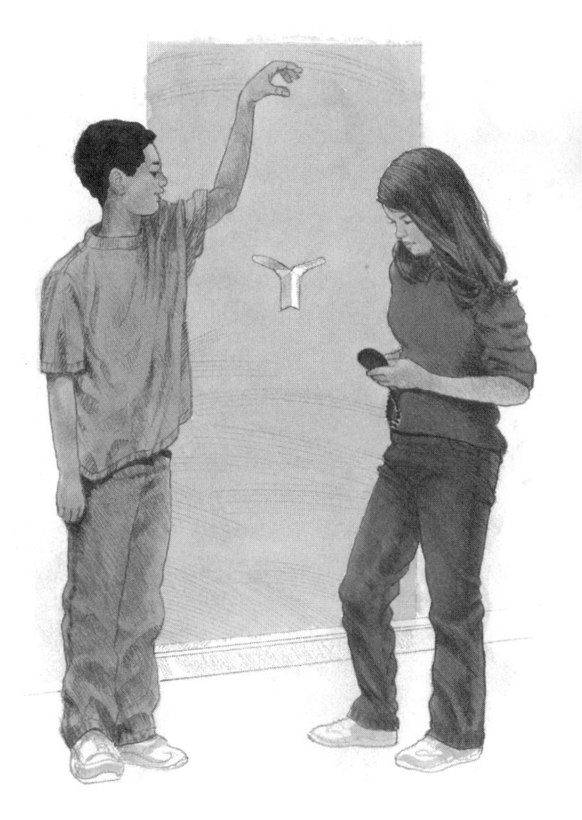

4.3 Explain and Recommend

4.3 Explain and Recommend

Explanations and Recommendations about Parachutes

As you did after your whirligig experiments, you will spend some time now explaining your results. You will also try to come up with recommendations. Remember that explanations include your claims, the evidence for your claims, and the science you know that can help you understand the claim. A recommendation is a statement about what someone should do. The best recommendations also have evidence, science, and an explanation associated with them. In the *Whirligig Challenge*, you created explanations and recommendations separately from each other. This time you will work on both at the same time.

Create and Share Your Recommendation and Explanation

Work with your group. Use the hints on the *Create Your Explanation* pages to make your first attempt at explaining your results. You'll read about parachute science later. After that, you will probably want to revise your explanations. Right now, use the science you learned during the *Whirligig Challenge* for your first attempt.

Write your recommendation. It should be about designing a slow-falling parachute. Remember that it should be written so that it will help someone else. They should be able to apply what you have learned about the effects of your variable. If you are having trouble, review the example in *Learning Set 3*.

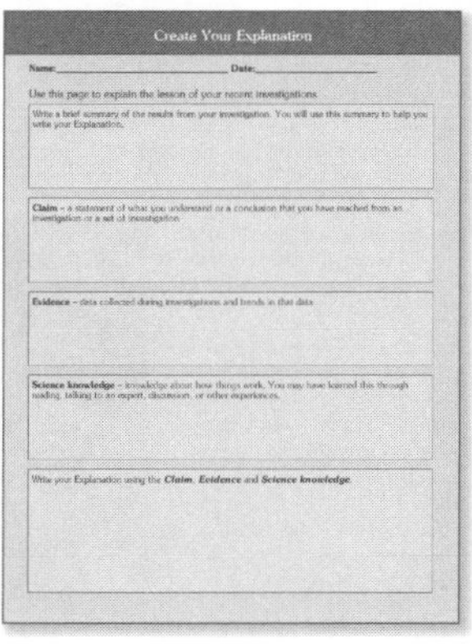
Create Your Explanation

Name:____________ Date:____________

Use this page to explain the lesson of your recent investigations.

Write a brief summary of the results from your investigation. You will use this summary to help you write your Explanation.

Claim – a statement of what you understand or a conclusion that you have reached from an investigation or a set of investigation.

Evidence – data collected during investigations and trends in that data.

Science knowledge – knowledge about how things work. You may have learned this through reading, talking to an expert, discussion, or other experiences.

Write your Explanation using the ***Claim***, ***Evidence*** and ***Science knowledge***.

81 DIVING IN TO SCIENCE

reservoir
drain hole
flat
slanted

Your teacher will set up the stream table in four different ways, as shown in the diagrams.

Sketch the different models. As you watch the water flow through the model, pay very close attention to the way the land on both sides of the river changes. Pay attention to

- how the soil moves,
- where along the bank the soil moves, and
- where the soil ends up.

Make notes about what you observe for each of these situations. You might want to mark your sketches based on what you observed.

Stop and Think

Look at your sketches and the notes you took about the river models you observed. What did you notice about how the soil was moved by the river? Answer these questions. Be prepared to discuss your answers with your group and the class.

1. When the river was straight and the pan was level, how did the soil move along the river?
2. When your teacher made the pan more slanted by lifting the water end of the pan, how did the water move compared to the level pan? How did that change affect the soil that the river moved?
3. Your teacher also made rivers that were more curved. How did that change the way the soil moved along the river?

Reflect

Think about the book support you designed and built so far. Try to think about the science concepts you have read about and discussed as a class. Answer the following questions. Be prepared to discuss your answers with the class.

1. Was your structure strong? If not, did it collapse because of folding, compression, or both?
2. How could you make the structure stronger to resist folding or compression?
3. Was your book support stable? That is, did it provide support so that the book did not tip over? Did it provide this support well? Draw a picture of your book support showing the center of mass of the book and the places in your book support that resist the load of your book.
4. How could you make your book support more stable?
5. How successful were the book supports that used columns in their design?
6. How could you make your book support work more effectively by including columns into the design?
7. Explain how the pull on the book could better be resisted by the use of columns in your design. Be sure to discuss both the strength and the stability of the columns in your design. You might find it easier to draw a sketch and label it to explain how the columns do this.
8. Think about some of the structures that supported the book well. What designs and building decisions were used?

You are going to get another chance to design a book support. You will use the same materials. Think about how your group could design your next book support to better meet the challenge. Consider what you now know about the science that explains how structures support objects.

3) Scientists...reflect in many different ways.

PBIS provides guidance to help students think about what they are doing and to recognize what they are learning. Doing this often as they are working will help students be successful student scientists.

Tools for Making Sense

Stop and Think

Stop and Think sections help students make sense of what they have been doing in the section they are working on. *Stop and Think* sections include a set of questions to help students understand what they have just read or done. Sometimes the questions will remind them of something they need to pay more attention to. Sometimes they will help students connect what they have just read to things they already know. When there is a *Stop and Think* in the text, students will work individually or with a partner to answer the questions, and then the whole class will discuss the answers.

Reflect

Reflect sections help students connect what they have just done with other things they have read or done earlier in the Unit (or in another Unit). When there is a *Reflect* in the text, students will work individually or with a partner or small group to answer the questions. Then, the whole class will discuss the answers. You may want to ask students to answer *Reflect* questions for homework.

Analyze Your Data

Whenever students have to analyze data, the text will provide hints about how to do that and what to look for.

Mess About

"Messing about" is a term that comes from design. It means exploring the materials to be used for designing or building something or examining something that works like what is to be designed. Messing about helps students discover new ideas—and it can be a lot of fun. The text will usually give them ideas about things to notice as they are messing about.

What's the Point?

At the end of each *Learning Set*, students will find a summary, called *What's the Point?*, of the important information from the *Learning Set*. These summaries can help students remember how what they did and learned is connected to the *Big Question* or *Challenge* they are working on.

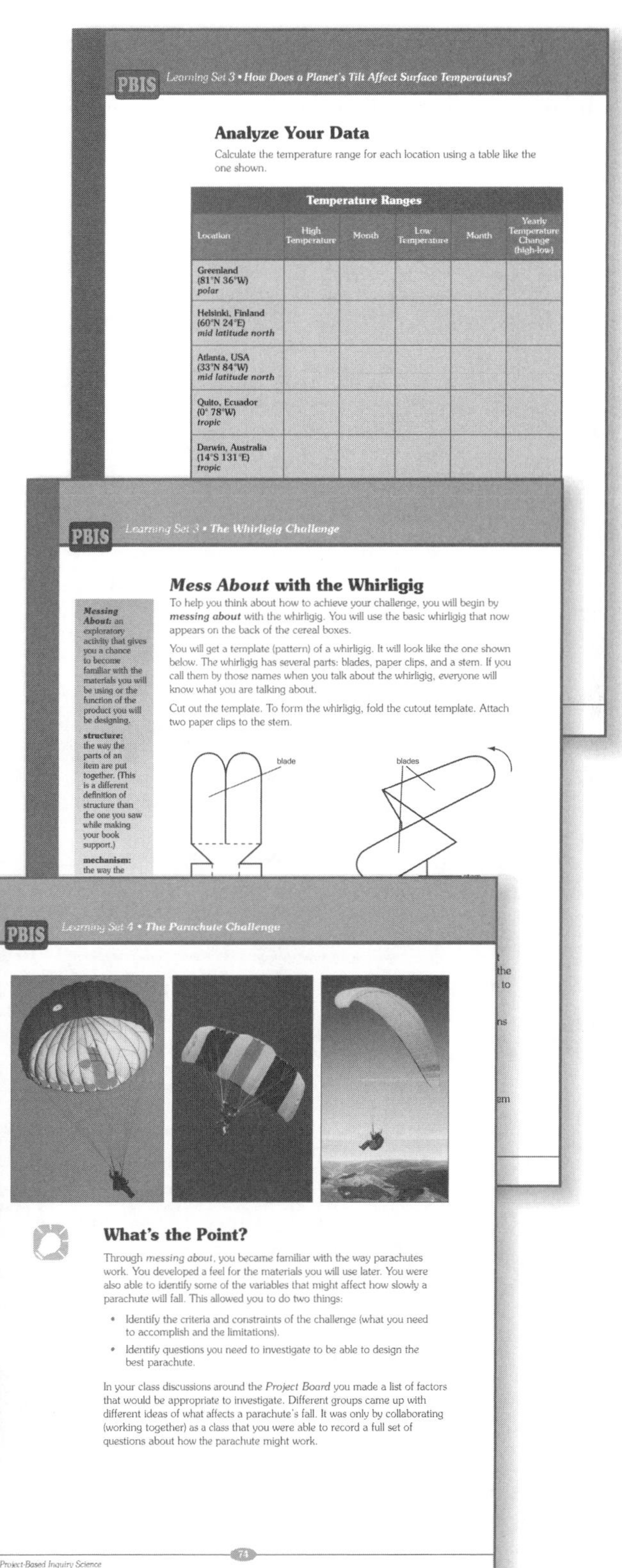

PBIS Learning Set 3 • How Does a Planet's Tilt Affect Surface Temperatures?

Analyze Your Data

Calculate the temperature range for each location using a table like the one shown.

Temperature Ranges					
Location	High Temperature	Month	Low Temperature	Month	Yearly Temperature Change (high-low)
Greenland (81°N 36°W) *polar*					
Helsinki, Finland (60°N 24°E) *mid latitude north*					
Atlanta, USA (33°N 84°W) *mid latitude north*					
Quito, Ecuador (0° 78°W) *tropic*					
Darwin, Australia (14°S 131°E) *tropic*					

PBIS Learning Set 3 • The Whirligig Challenge

***Mess About* with the Whirligig**

To help you think about how to achieve your challenge, you will begin by ***messing about*** with the whirligig. You will use the basic whirligig that now appears on the back of the cereal boxes.

You will get a template (pattern) of a whirligig. It will look like the one shown below. The whirligig has several parts: blades, paper clips, and a stem. If you call them by those names when you talk about the whirligig, everyone will know what you are talking about.

Cut out the template. To form the whirligig, fold the cutout template. Attach two paper clips to the stem.

Messing About: an exploratory activity that gives you a chance to become familiar with the materials you will be using or the function of the product you will be designing.

structure: the way the parts of an item are put together. (This is a different definition of structure than the one you saw while making your book support.)

mechanism: the way the

PBIS Learning Set 4 • The Parachute Challenge

What's the Point?

Through *messing about*, you became familiar with the way parachutes work. You developed a feel for the materials you will use later. You were also able to identify some of the variables that might affect how slowly a parachute will fall. This allowed you to do two things:

- Identify the criteria and constraints of the challenge (what you need to accomplish and the limitations).
- Identify questions you need to investigate to be able to design the best parachute.

In your class discussions around the *Project Board* you made a list of factors that would be appropriate to investigate. Different groups came up with different ideas of what affects a parachute's fall. It was only by collaborating (working together) as a class that you were able to record a full set of questions about how the parachute might work.

Project-Based Inquiry Science 74

PBIS Learning Set 3 • How Can Changes in Water Quality Affect the Living Things in an Ecosystem?

3.3 Explore

The Marry Martans River Mystery: Macroinvertebrates in an Ecosystem

ecologist: a scientist who studies the relationships between organisms and their environment.

You watched a video of scientists collecting macroinvertebrates. You should now have a good sense of how scientists organize and classify macroinvertebrates. Once scientists identify macroinvertebrates in an ecosystem, they can use this information to better understand the conditions in an ecosystem.

You also learned about diversity and abundance. Recall that diversity refers to the types of organisms found in an environment. Abundance refers to the number of each type. In this activity, you will examine the diversity of macroinvertebrates in an area. You will see how diversity can indicate water quality and ecosystem health. You will be working with some macroinvertebrate data collected by an **ecologist**. The ecologist has been asked to help the residents of a small community solve a mystery. What you learn from this case study will help you address this *Learning Set's* question.

Examine a Case Study

A group of residents live on a small lake called Marry Martans Lake. The Marry Martans River flows into the lake at one end. The lake drains back into the river at the other end. (See the picture on the next page.) Over the past few months, the residents have noticed a lot of algae growing in the lake. The young people in the community know about water-quality indicators from their science classes. They remember that sudden algae and plant growth could be a sign of high amounts of fertilizer running off into the river.

The young people and their parents decide to investigate the case. Where might the fertilizer be coming from? They discover that there are three farms upriver. These farms are upstream from the lake and border the river.

PBIS Learning Set 1 • How Do Flowing Water and Land Interact in a Community?

Communicate Your Results

Investigation Expo

Scientists always share their understandings with each other. Presenting their results to others is one of the most important things that scientists do. You will share what you have found in an *Investigation Expo*. To prepare for this, you will use an overhead transparency.

Trace the diagram that you drew of your model onto an overhead transparency. Be ready to describe your investigation and clearly detail all your results. The answers to the following questions will be very helpful in preparing your presentation.

- Describe how the water moved in the model. (What patterns did you see?)
- Describe why you think your prediction was accurate or inaccurate.
- How did the outcome compare to your prediction?
- Where did the water flow more quickly? How was the flow different from what you predicted?
- Where did the water pool?

As you look at the overheads presented by other students, make sure you can answer these questions. Ask questions that you need answered to understand the results and the explanations others have made.

What's the Point?

In this section, you built a model to simulate how water flows across a landscape when it rains. Placing different-sized objects under the paper created the higher and lower elevations. You also drew a sketch of the model, and predicted how water will run over the paper. When you ran the simulation you probably noticed that the water always moved from areas of high elevation to areas of lower elevation. Water cannot move uphill. You also noticed that water flowed and created puddles in several places as it flowed. These puddles represent lakes or ponds in the real world.

Water on land works that way too. If you watch where rain falls in one rainstorm, you will be able to predict the path water will take in the next rainstorm. This is the case as long as the land stays the same from the first rainstorm to the next. You will need to consider how new construction in Wamego might change the land and affect how water flows.

Project-Based Inquiry Science 20

4) Scientists...collaborate.

Scientists never do all their work alone. They work with other scientists (collaborate) and share their knowledge. *PBIS* helps students by giving them lots of opportunities for sharing their findings, ideas, and discoveries with others (the way scientists do). Students will work together in small groups to investigate, design, explain, and do other science activities. Sometimes they will work in pairs to figure out things together. They will also have lots of opportunities to share their findings with the rest of their classmates and make sense together of what they are learning.

Investigation Expo

In an *Investigation Expo*, small groups report to the class about an investigation they've done. For each *Investigation Expo*, students will make a poster detailing what they were trying to learn from their investigation, what they did, their data, and their interpretation of the data. The text gives them hints about what to present and what to look for in other groups' presentations. *Investigation Expos* are always followed by discussions about the investigations and about how to do science well. You may want to ask students to write a lab report following an investigation.

Plan Briefing/Solution Briefing/ Idea Briefing

Briefings are presentations of work in progress. They give students a chance to get advice from their classmates that can help them move forward. During a *Plan Briefing*, students present their plans to the class. They might be plans for an experiment for solving a problem or achieving a challenge. During a *Solution Briefing*, students present their solutions in progress and ask the class to help them make their solutions better. During an *Idea Briefing*, students present their ideas, including their evidence in support of their plans, solutions, or ideas. Often, they will prepare posters to help them make their presentation. Briefings are almost always followed by discussions of their investigations and how they will move forward.

Solution Showcase

Solution Showcases usually happen near the end of a Unit. During a *Solution Showcase*, students show their classmates their finished products—either their answer to a question or solution to a challenge. Students will also tell the class why they think it is a good answer or solution, what evidence and science they used to get to their solution, and what they tried along the way before getting to their answers or solutions. Sometimes a *Solution Showcase* is followed by a competition. It is almost always followed by a discussion comparing and contrasting the different answers and solutions groups have come up with. You may want to ask students to write a report or paper following a *Solution Showcase*.

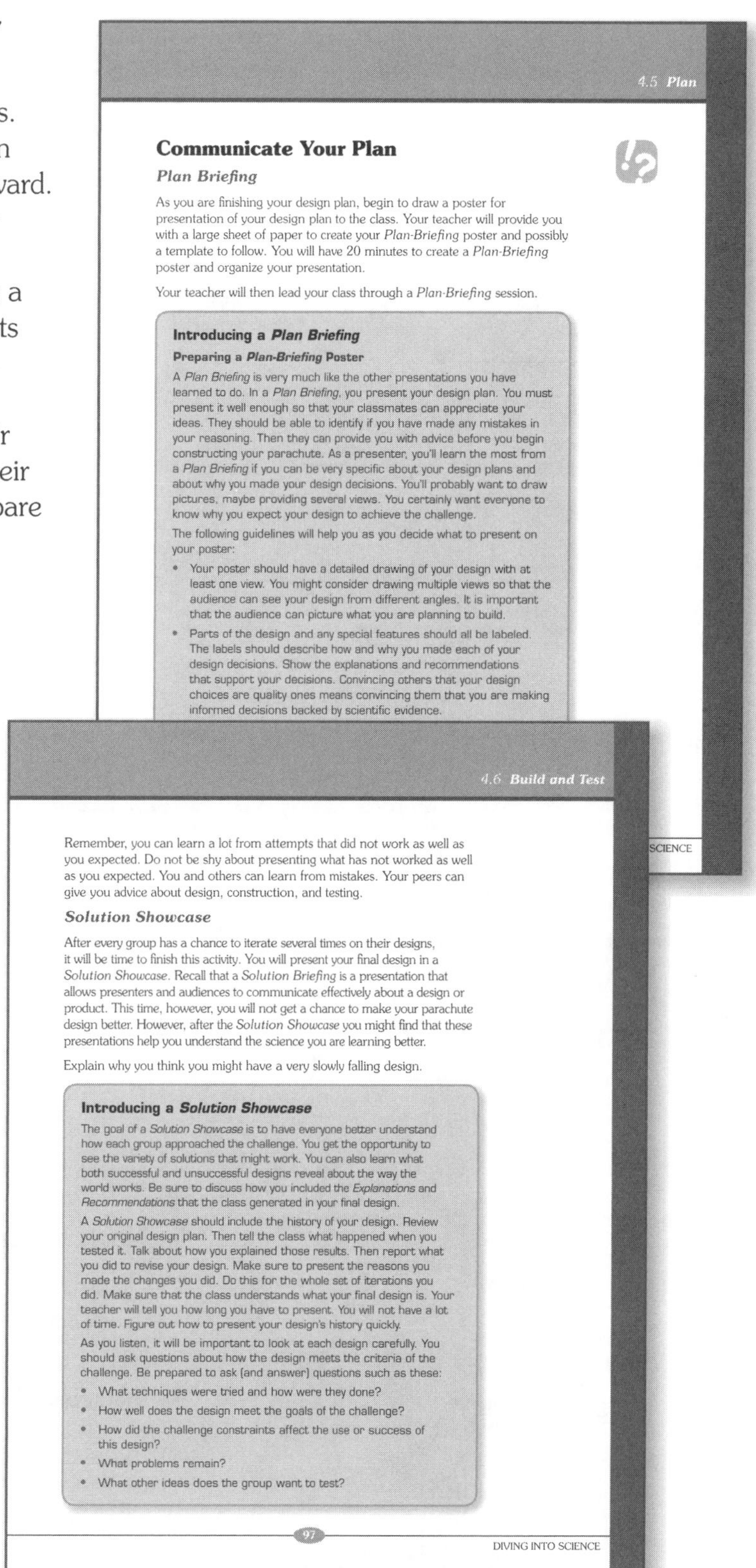

4.5 *Plan*

Communicate Your Plan

Plan Briefing

As you are finishing your design plan, begin to draw a poster for presentation of your design plan to the class. Your teacher will provide you with a large sheet of paper to create your *Plan-Briefing* poster and possibly a template to follow. You will have 20 minutes to create a *Plan-Briefing* poster and organize your presentation.

Your teacher will then lead your class through a *Plan-Briefing* session.

Introducing a *Plan Briefing*

Preparing a *Plan-Briefing* Poster

A *Plan Briefing* is very much like the other presentations you have learned to do. In a *Plan Briefing*, you present your design plan. You must present it well enough so that your classmates can appreciate your ideas. They should be able to identify if you have made any mistakes in your reasoning. Then they can provide you with advice before you begin constructing your parachute. As a presenter, you'll learn the most from a *Plan Briefing* if you can be very specific about your design plans and about why you made your design decisions. You'll probably want to draw pictures, maybe providing several views. You certainly want everyone to know why you expect your design to achieve the challenge.

The following guidelines will help you as you decide what to present on your poster:

- Your poster should have a detailed drawing of your design with at least one view. You might consider drawing multiple views so that the audience can see your design from different angles. It is important that the audience can picture what you are planning to build.
- Parts of the design and any special features should all be labeled. The labels should describe how and why you made each of your design decisions. Show the explanations and recommendations that support your decisions. Convincing others that your design choices are quality ones means convincing them that you are making informed decisions backed by scientific evidence.

4.6 *Build and Test*

Remember, you can learn a lot from attempts that did not work as well as you expected. Do not be shy about presenting what has not worked as well as you expected. You and others can learn from mistakes. Your peers can give you advice about design, construction, and testing.

Solution Showcase

After every group has a chance to iterate several times on their designs, it will be time to finish this activity. You will present your final design in a *Solution Showcase*. Recall that a *Solution Briefing* is a presentation that allows presenters and audiences to communicate effectively about a design or product. This time, however, you will not get a chance to make your parachute design better. However, after the *Solution Showcase* you might find that these presentations help you understand the science you are learning better.

Explain why you think you might have a very slowly falling design.

Introducing a *Solution Showcase*

The goal of a *Solution Showcase* is to have everyone better understand how each group approached the challenge. You get the opportunity to see the variety of solutions that might work. You can also learn what both successful and unsuccessful designs reveal about the way the world works. Be sure to discuss how you included the *Explanations* and *Recommendations* that the class generated in your final design.

A *Solution Showcase* should include the history of your design. Review your original design plan. Then tell the class what happened when you tested it. Talk about how you explained those results. Then report what you did to revise your design. Make sure to present the reasons you made the changes you did. Do this for the whole set of iterations you did. Make sure that the class understands what your final design is. Your teacher will tell you how long you have to present. You will not have a lot of time. Figure out how to present your design's history quickly.

As you listen, it will be important to look at each design carefully. You should ask questions about how the design meets the criteria of the challenge. Be prepared to ask (and answer) questions such as these:

- What techniques were tried and how were they done?
- How well does the design meet the goals of the challenge?
- How did the challenge constraints affect the use or success of this design?
- What problems remain?
- What other ideas does the group want to test?

97

DIVING INTO SCIENCE

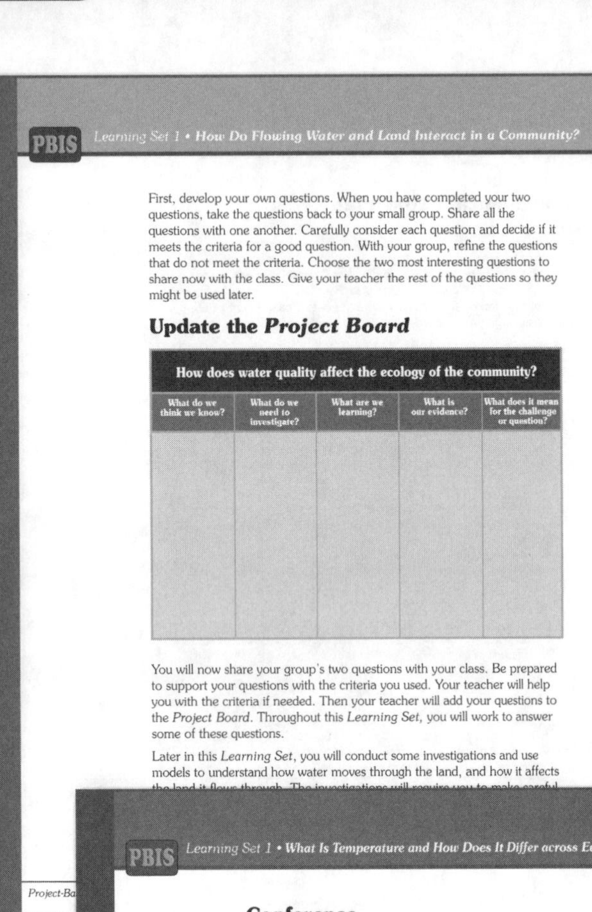
PBIS Learning Set 1 • How Do Flowing Water and Land Interact in a Community?

First, develop your own questions. When you have completed your two questions, take the questions back to your small group. Share all the questions with one another. Carefully consider each question and decide if it meets the criteria for a good question. With your group, refine the questions that do not meet the criteria. Choose the two most interesting questions to share now with the class. Give your teacher the rest of the questions so they might be used later.

Update the *Project Board*

How does water quality affect the ecology of the community?				
What do we think we know?	What do we need to investigate?	What are we learning?	What is our evidence?	What does it mean for the challenge or question?

You will now share your group's two questions with your class. Be prepared to support your questions with the criteria you used. Your teacher will help you with the criteria if needed. Then your teacher will add your questions to the *Project Board*. Throughout this *Learning Set*, you will work to answer some of these questions.

Later in this *Learning Set*, you will conduct some investigations and use models to understand how water moves through the land, and how it affects

Project-Ba

PBIS Learning Set 1 • What Is Temperature and How Does It Differ across Earth's Surface?

Conference

Teams of scientists often work together to solve problems. They hold group discussions. That is what you are going to do. During your discussion, you can present questions that you have. Sometimes if you do not have an answer, someone else might. You might also present a question that no one else had thought of. This can start your group thinking in a new direction.

Discuss your map with a partner and then with your group. Listen and observe as others present their maps to the group. As you present your prediction map, include answers to these questions:

- How did you decide what temperatures to use to color each area?
- How did you decide where to start and where to go to next?
- In which parts of the world do you feel very confident about your predictions, and which parts do you feel unsure about?

After everyone has presented their maps, take note of where there was agreement and where there were differences. Later on you will compare your predictions to a real surface-temperature map.

You have compared your temperature predictions for Earth with those of others in your group. Now, work again with your partner to create a prediction map based on discussions you've just had. Begin with a... group were in agreement. ... is disagreement ... few minutes t... or evidence. ... something, t...

Project-Based Inquiry Science PF 18

Update the *Project Board*

Remember that the *Project Board* is designed to help the class keep track of what they are learning and their progress toward a Unit's *Big Question* or *Challenge*. At the beginning of each Unit, the class creates a *Project Board*, and together records what students think they know about answering the *Big Question* or addressing the *Big Challenge* and what they think they need to investigate further. Near the beginning of each *Learning Set*, the class revisits the *Project Board* and adds new questions and information they think they know to the *Project Board*. At the end of each *Learning Set*, the class again revisits the *Project Board*. This time, they record what they have learned, the evidence they have collected, and recommendations they can make about answering the *Big Question* or achieving the *Big Challenge*.

Conference

A *Conference* is a short discussion among a small group of students before a more formal whole-class discussion. Students might discuss predictions and observations, they might try to explain together, they might consult on what they think they know, and so on. Usually, a *Conference* is followed by a discussion around the *Project Board*. In these small group discussions, everybody gets a chance to participate.

What's the Point?
Students review what they have learned in each *Learning Set*.

Stop and Think
Student answer questions that help them understand what they have done in a section.

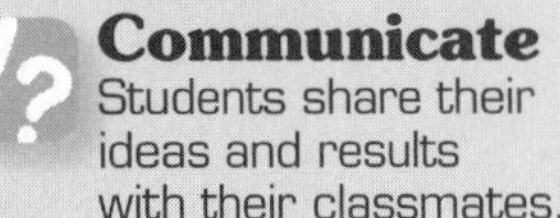
Communicate
Students share their ideas and results with their classmates.

Record
Students record their data as they gather it.

NOTES

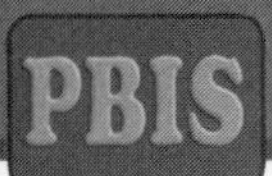

NOTES

NOTES

NOTES

PBIS
DIGGING IN
As a Student Scientist, you will...
Ask QUESTIONS
Pursue ANSWERS
APPLY MEANING
Make MEANING
Share ANSWERS
Teacher's Planning Guide

UNIT OVERVIEW

Launcher Units are designed to introduce students to the practices of science and social practices of the classroom. As students engage in the social practices of scientists they learn what scientists do and how they do it, and begin to think of themselves as student scientists. The Launcher Unit provides anchoring experiences to refer to throughout the year. It is also where you, the teacher, set expectations for the year about rigor and ways of working together. During this Unit, remember to emphasize collaboration, analyzing data, designing investigations, and constructing explanations.

Content

This Launcher Unit builds experience with *Project Based Inquiry Science* learning and fosters a culture of collaboration and rigorous scientific discourse in the classroom. In this Unit, the *Big Question* asks, *How do scientists work together to solve problems?* Within the context of earth science, students explore variables, reliable procedures, fair tests, case studies, and the use of evidence to construct explanations and make recommendations. Throughout the Unit, students engage in the social practices of scientists as they address a set of small challenges. The goals of *Learning Set 1* and *Learning Set 2* are focused primarily on the social practices of scientists and the investigative process, including iteration, collaboration, measuring, running an experiment, analyzing experimental data, and getting reliability results. In *Learning Set 3,* students have an opportunity to refine their understanding and use of the whole set of practices and concepts they have learned in the context of designing an erosion-control method. They continue practicing as scientists while designing their first model and simulation, constructing scientific explanations, and applying the science they are learning to make recommendations. To conclude the Unit, students watch a video of a design team using all the practices the students have been learning to design a new type of shopping cart. Students see the value of the collaboration and scientific reasoning practices in a real-world situation. They use this as a context to pull together what they have learned to answer the *Big Question.*

LOOKING AHEAD

Many of the practices of scientists and social practices of *PBIS* are introduced to students in this Unit. There is information on each in the *Teacher's Resource Guide.*

Investigations

Student groups have experiences that lead them to answers in the Unit's *Big Question.* In *Learning Set 1,* students learn the collaborative nature of science as they plan, build, and test a boat that is able to carry a cargo of keys with their group and hear about other groups' designs during class presentations. Students learn the importance of criteria and constraints of design and the value of the iterative process as they redesign and build their boats. Through group

28 class periods

A class period is considered to be one 40 to 50 minute class.

investigations of how fast model lava flows across a plate in *Learning Set 2*, students realize the need for a standard set of procedures and the requirement of reproducible results for an investigation to be considered valid. Within this context, they learn about errors and measurement constraints. Students practice graphing and analyzing data. Students engage in scientific social practices as their groups present results to the class and discuss them. While building a method for erosion control in *Learning Set 3*, groups apply these skills as they design models and simulations, along with investigations to test their designs. They design their erosion-control methods based on evidence they have collected through case studies and investigations, as well as scientific knowledge. They continue practicing collaboration, investigative design, constructing explanations, and making recommendations while designing, building, and testing their erosion-control methods.

Nature of Science

Through practice, students learn how scientists refine their ideas as new information becomes available and learn how scientists work together to refine their understanding. They use what they know to answer important questions. Through an iterative process of designing and building a boat several times, students learn how scientific understanding can help them develop good solutions to everyday challenges. Students learn that knowledge is shared and used by the rest of the science community, their class in this case, and the importance of giving credit to the originator of an idea or invention. While engaging in the social practices of scientists, students realize the value of sharing and building on each other's ideas. While conducting investigations in *Learning Set 2*, students realize that for results to be considered valid, procedures must be replicable with repeatable results. This requires that procedures are communicated clearly and specifically so others can follow them. To help students reflect on how their ideas change in the same way scientists' ideas do, they are introduced to the *Project Board* in *Learning Set 3*, a visual tool for keeping a running record of what they know, what they need to investigate, their claims with the evidence they have to back these up, as well as how all the information fits together to answer a bigger question.

Artifacts

In *Learning Set 1*, student groups design and build a boat, satisfying a number of criteria and constraints that are given to them, as well as some that the class decides upon. Students keep records of their design iterations and the designs of others. In *Learning Set 2*, students write procedures for using models and simulations, record their data, and use a graph of the class's data. In *Learning Set 3*, students design and build erosion control methods. The class uses a *Project Board* to keep track of the different causes of erosion and different methods that have been used by others to control erosion. Student groups use models and simulations to test the effectiveness of their erosion control method.

They design investigations, gather data, and interpret their results. They also write letters of recommendation to a school board about how to control erosion at a proposed basketball court site.

Targeted Concepts, Skills, and Nature of Science	Section
Scientists often work together and then share their findings. Sharing findings makes new information available and helps scientists refine their ideas and build on others' ideas. When another person's or group's idea is used, credit needs to be given.	1.1, 1.2, 1.3, 1.4, 2.1, 2.3, 2.4, 3.1, 3.2, 3.3, 3.4, 3.5, 3.6, 3.7, 3.8, 3.9, 3.10
Criteria and constraints are important in design.	1.1,1.2, 1.4, 3.1, 3.6
Scientists must keep clear, accurate, and descriptive records of what they do so they can share their work with others and consider what they did, why they did it, and what they want to do next.	1.2, 1.4, 2.2, 3.1, 3.2, 3.3, 3.6, 3.7, 3.8, 3.9, 3.10
Graphs are an effective way to communicate results of scientific investigation.	2.2
Identifying factors that could affect the results of an investigation is an important part of planning scientific research.	2.3, 3.2, 3.3, 3.5, 3.6
Scientific investigations and measurements are considered reliable if the results are repeatable by other scientists using the same procedures.	2.2, 2.3, 2.4, 3.3, 3.6, 3.7
Science and engineering are dynamic processes, changing as new information becomes available.	1.3,1.4
In a fair test only the manipulated (independent) variable, and the responding (dependent) variable change. All other variables are held constant.	3.3, 3.6, 3.7
Scientists use models to simulate processes that happen too fast, too slow, on a scale that cannot be observed directly (either too small or too large), or that are too dangerous.	2.1, 2.2, 3.6, 3.7, 3.9, 3.10
Scientists make claims (conclusions) based on evidence obtained (trends in data) from reliable investigations.	3.2, 3.3, 3.4, 3.7, 3.10
Explanations are claims supported by evidence, accepted ideas and facts.	3.4, 3.8, 3.10
Earth's gravity pulls things toward Earth.	1.3

Targeted Concepts, Skills, and Nature of Science	Section
When an object is immersed in a liquid it experiences an upward force on it called the buoyant force. (which is equal to the weight of the amount of fluid the object has displaced).	1.3
Density is the mass per volume of an object.	1.3
The density and surface area of an object determine if the object will sink or float.	1.3
When volcanoes erupt, magma reaches Earth's surface and is called lava. There are many different types of lava.	2.1, 2.4
Erosion is the process of soil and other particles being displaced by water, waves, wind, and gravity.	3.1, 3.2, 3.5, 3.6, 3.7, 3.8, 3.9, 3.10

NOTES

Unit Materials List

Quantities for groups of 4-6 students.		
Unit Durable Group Items	**Section**	**Quantity**
Hex nut 1/2"	1.1, 1.2, 1.4	4
Scissors	1.1, 1.2, 1.4	1
Plastic shoebox, for stream table	1.1, 1.2, 1.4, 3.3, 3.6, 3.7, 3.10	1
Stream table apparatus	1.1, 1.2, 1.4, 3.3, 3.6, 3.7, 3.10	1
Stopwatch	1.2, 1.4, 2.2, 2.4, 3.3	1
Key blank	1.4	2
Bottle, 4 oz	2.2, 2.4	1
Ruler, in./cm	2.2, 2.4	1
Measuring spoon set	2.2, 2.4	1
Plastic plate, 9" diameter	2.2, 2.4	1
Plastic cup, 10 oz	3.3, 3.6, 3.7, 3.10	1

NOTES

Quantities for 5 classes of 8 groups.		
Unit Durable Classroom Items	**Section**	**Quantity**
Project Board, laminated	3.1, 3.2, 3.5, 3.7, 3.8, 3.10, ABC	5
Project Board transparency	3.1, 3.2, 3.5, 3.7, 3.8, 3.10, ABC	1
Measuring cup	3.3	2
Plastic Drop Cloth, approx. 6' x 4' for the stream table	3.3, 3.6, 3.7, 3.10	4
Spray bottle	3.3, 3.6, 3.7, 3.10	4
IDEO Deep Dive DVD	ABC	1

Quantities for 5 classes of 8 groups.		
Consumable Classroom Items	**Section**	**Quantity**
Heavy-duty foil, roll	1.1, 1.2, 1.3, 1.4	1
Masking tape	1.1, 1.2, 1.3, 1.4	1
Liquid soap	2.2, 2.4	2
Graph paper, pkg. of 50	2.2, 2.4	1
Restickable easel pad	3.2, 3.3, 3.7, ABC	2
Colored pencils	3.2, 3.3, 3.7, ABC	6

Unit Materials List

Quantities for 5 classes of 8 groups.		
Consumable Classroom Items	**Section**	**Quantity**
Sand, fine grain, 5 lb	3.3	3
Disposable gloves, pkg. of 100	3.3, 3.6, 3.7, 3.10	2
Landscape Materials, 5 lb	3.3, 3.7, 3.10	2
Potting soil, 5 lb	3.6	2
Spanish Moss, 10" x 13" bag	3.6, 3.7, 3.10	1
Slate chips, 1 lb	3.7, 3.10	2

Additional Items Needed Not Supplied	Section	Quantity
Access to water	1.1. 1.2, 1.4, 3.3, 3.6, 3.7, 3.10	1 per classroom
Paper towel, roll	1.1. 1.2, 1.4, 3.3, 3.6, 3.7, 3.10	1 per classroom
Access to outdoors to conduct an erosion walk	3.1	1 per classroom
Books, to prop up stream table	3.3, 3.6, 3.7, 3.10	as needed
Native soil sample	3.3, 3.7, 3.10	2 cups per group

UNIT INTRODUCTION

What's the Big Question?

How Do Scientists Work Together to Solve Problems?

Overview

Students learn that the work of scientists involves problem-solving skills, and they are introduced to the *Big Question* of the Unit, How do scientists work together to solve problems? They learn they will work on three challenges in this Unit, solving problems and having experiences that will require them to use many of the practices of scientists. At the end of the Unit, they will use what they have learned from their experiences to answer the *Big Question.*

Targeted Concepts, Skills, and Nature of Science	Performance Expectations
Scientists often work together and then share their findings. Sharing findings makes new information available and helps scientists refine their ideas and build on others' ideas. When another person or group's idea is used, credit needs to be given.	Students should demonstrate understanding of the *Big Question*.

Homework Options

Reflection

- **Science Process:** How do you think the skills that scientists use could be useful in everyday life? When might you use these skills? Describe a situation when you could use them. *(Students should describe problems in everyday life that they could solve using analytical and logical thinking.)*

NOTES

UNIT INTRODUCTION IMPLEMENTATION

What's the Big Question?

How Do Scientists Work Together to Solve Problems?

Welcome to your new science Unit. This Unit and others you complete this year will offer you exciting challenges and opportunities to learn science. Science involves learning very interesting facts, but that's not all science is. A large part of learning science is being able to analyze and make sense of the world around you in an organized and logical way. Scientists learn how to do this to be successful at what they do. But scientists are not the only ones who can benefit from this kind of reasoning; you may also find it useful.

In this Unit, you will learn how to tackle problems and challenges as a scientist does. There is a lot scientists do to make sure that they solve problems in an organized and logical way. You will experience and use many of these scientific practices. You are not expected to learn everything about being a scientist in just a few weeks, but you will learn a lot. The lessons you learn will help you be successful in science class this year. They will also help you in future science classes and even in your life!

Your *Big Question* in this Unit is to understand how scientists solve problems. To help you answer the *Big Question*, you will work on three challenges in this Unit. Each one will give you a chance to learn a few practices and behaviors. Then you will use what you have learned to answer the *Big Question*: *How do scientists work together to solve problems?*

Some of you may have already begun learning about what scientists do and how they work together. For you, much of this Unit will be review. That review will be useful to you. It will give the members of your class a chance to learn to work together. It will also allow you to share what you have learned in other years with your new classmates. You will also find new things to learn in this Unit. This Unit introduces new science content and practices of scientists that you have not discussed a lot before.

Have fun being student scientists!

DIG 3

DIGGING IN

What's the Big Question?

How Do Scientists Work Together to Solve Problems?

5 min.

Students are introduce to PBIS *and the* Big Question *they will be attempting to answer throughout this Unit.*

META NOTES

Many students have narrow ideas about who scientists are and what scientists do. They may think that scientists work exclusively in laboratories, or that they only answer arcane or academic questions. They do not see themselves as scientists, and they may not realize that science is often about answering big questions and applying what is known to addressing real-world challenges.

○ Engage

Begin by eliciting students' ideas about what scientists do. Ask students if they have ever done anything the way scientists do. Some students may explain something they know from science they have read. This is scientific knowledge, and using scientific knowledge is an important part of science. Try to elicit ideas about solving problems. Ask students if they have ever solved a difficult problem or ever watched anyone solve a difficult problem. Let them know that analyzing and making sense of the world in a logical way, as you do when you solve a problem, is an important part of science.

TEACHER TALK

"What do scientists do? How do they do that? Have you ever done things the way scientists do? Have you seen anybody do things the way scientists do? One important part of science is analyzing and making sense of the world in a logical way. This helps you to solve difficult problems."

Get Going

Read *What's the Big Question?* with the class. Emphasize that students will learn and use the skills that scientists use as they work through the Unit.

Teacher Reflection Questions

- What misconceptions did you find in students' responses to your questions about what scientists do? How will these misconceptions be challenged in this Unit? What can you do to challenge these misconceptions?
- How can you engage students in learning about the practices of scientists?
- How did students contribute to the discussion when you asked them questions? How can you encourage students' participation? How can you evaluate students' participation?

NOTES

LEARNING SET 1 INTRODUCTION

Learning Set 1

The Build a Boat Challenge

◀ ***5 class periods***

A class period is considered to be one 40 to 50 minute class.

Thinking about criteria and constraints, student groups build and test a boat made out of a piece of aluminum foil that must support six keys.

Overview

After students are given the *Big Question: How do scientists work together to solve problems?* and the *Learning Set* challenge, they are introduced to the concept of criteria and constraints to help them plan their solutions to the challenge. The challenge is to build a boat that can carry six keys for at least 20 seconds using only a 5" X 5" piece of foil. Students begin by quickly identifying the specific criteria and constraints of the challenge, planning and assembling a design solution for a boat, and then presenting their solutions to their peers. When groups build their boats, most will have different designs because they interpreted the requirements differently. Based on the experience of sharing boat designs and gathering ideas from other groups, students update their criteria and constraints. Groups redesign and rebuild their boats, applying two common practices of science: the iterative process and building on others' ideas. Groups present their new designs to the class in their first *Solution Briefing,* learning the importance of crediting other people's work. After this *Solution Briefing,* students read about the science of boat design and use new science knowledge to redesign and rebuild their boat with the new criterion, it now must carry eight keys. At the end of the *Learning Set* they examine the practices of scientists that they have been engaging in.

Targeted Concepts, Skills, and Nature of Science	Section
Scientists often work together and then share their findings. Sharing findings makes new information available and helps scientists refine their ideas and build on others' ideas. When another person's or group's idea is used, credit needs to be given.	1.1, 1.2, 1.3, 1.4
Criteria and constraints are important in design.	1.1,1.2, 1.4

Targeted Concepts, Skills, and Nature of Science	Section
Scientists must keep clear, accurate, and descriptive records of what they do so they can share their work with others, consider what they did, why they did it, and what they want to do next.	1.2, 1.4
Earth's gravity pulls things toward Earth.	1.3
When an object is immersed in a liquid it experiences an upward force on it called the buoyant force (which is equal to the weight of the amount of fluid the object has displaced).	1.3
Density is the mass per volume of an object.	1.3

Students' Initial Conceptions and Capabilities

- Students generally believe that scientific knowledge changes, but they may think it changes only in facts and technology without realizing that a change in knowledge may suggest a reinterpretation of observations or new predictions. (Aikenhead, 1987; Lederman & O'Malley, 1990; Waterman, 1983.)
- Middle school students generally do not differentiate the concepts of mass, volume, and density and tend to use words such as "lighter," "heavier," "smaller," and "bigger" when describing objects. (Lee et. al, 1993; Stavy, 1990; Carey, 1991; Smith et. al, 1985; Smith, Snir, & Grosslight, 1987.)
- A common confusion between mass and density occurs because many believe that air (or gas) is weightless, and that adding air makes something lighter and taking away makes it heavier. (Lee et. al, 1993; Stavy, 1990.)
- Some students may believe that liquids and gases are weightless. (Stavy, 1991; Mas, Perez, and Harris, 1987.)

NOTES

Understanding for Teachers

The goal of this Unit is to familiarize students with the nature of science, the *PBIS* curriculum, and some of the tools used in this curriculum. It is not necessary for students to gain a deep understanding of the scientific concepts of mass, gravity, buoyancy, density, or balanced forces. However, students will need to apply these concepts to achieve their challenge. Although information is provided for the students to achieve the goal, some students may have more questions. In order for you to address any questions that arise, we provide more detail of the science concepts in this segment.

Matter

All matter is made up of atoms. This includes solids, liquids, and gases. Atoms are composed of a nucleus containing protons and neutrons (these are made up of quarks). Electrons reside outside of the nucleus. The particles that make up an atom have mass.

Gravity

All mass attracts mass. This is the source of gravity. Two objects with mass will pull on each other—this is gravitational interaction. The objects do not have to be touching to feel a pull from the other object (e.g., the Sun and Earth pull on each other even though they do not touch.) If one of the objects has a lot of mass (e.g., Earth) then the pull between the two objects interacting is strong. This is why things fall to Earth, the Moon orbits around Earth and Earth orbits around the Sun.

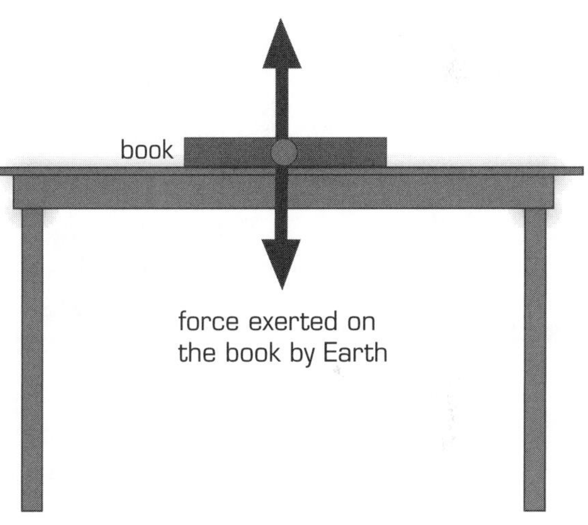

An object will not fall to Earth if there is a force equal in size but opposite in direction to the gravitational pull on it. An object sitting on a table will feel the gravitational pull exerted down on it by Earth, and will feel an equal but opposite force from the table pushing up on it. The force from the table is often called the normal force and arises because of the contact between the book and the table.

The normal force on the macroscopic level is a compression force. It is always perpendicular to the contact surface and pointing away from it. On the molecular scale, electromagnetic interactions happen between the particles that make up the surfaces in contact with each other.

When an object floats or is suspended in a fluid, the gravitational force exerted on it by Earth is being balanced by a force known as the buoyant force that is equal in size to the gravitational force acting on the object, but opposite in direction.

Density

Density is a measure of how much mass is in a given volume. It is a characteristic property, which helps scientists determine what something is made of. Each element has a unique density. The density of an object can be calculated using the formula:

$$Density = \frac{Mass}{Volume}$$

In general, an object will float if its density is less than the density of the fluid it is placed in. It will sink if its density is greater than the density of the fluid it is placed in. And it will remain where it is if its density is the same as the density of the fluid it is placed in. This is because of the opposing actions of the buoyant force and gravity on the object.

Buoyancy

The buoyant force arises from the pressure exerted by a fluid on an object that is in the fluid (liquid or gas). When an object is immersed in a liquid, a force from the liquid acts over the object's entire surface. This force is the pressure on the object from the liquid. If the fluid is not moving, the pressure is calculated using the following equation:

pressure = density X acceleration due to gravity X height below the surface

The buoyant force is equal in magnitude to the pressure times the surface area of contact. The force is perpendicular to the surface. Because it is not equal on all sides of an object, the overall force points upward. This upward force opposes the gravitational pull of Earth and, if it is strong enough, will cause the object to float.

The buoyant force exerted on an object by a fluid is equal to the weight of the fluid displaced by the object. This is the same as Archimedes' Principle — the net upward push exerted on an object partially or fully immersed in a fluid is equal in magnitude to the weight of the fluid displaced.

When the buoyant force exerted on an object by a fluid is greater than the gravitational force exerted on the object by Earth, the object rises in the fluid. When it is equal to the gravitational force exerted on the object by Earth, the object floats. When it is less than the gravitational force exerted on the object by Earth, the object sinks.

LEARNING SET 1 IMPLEMENTATION

Learning Set 1

The Build a Boat Challenge

Imagine that you and your friends are visiting an historic fort in a nearby state park. It is getting dark, and the park and fort will be closing soon. You and four friends walk over the drawbridge just as the gatekeeper is locking the door and preparing to raise the bridge. When you get to the parking lot, you realize that two of your friends are missing!

You rush back to the gate, but it is locked, and the drawbridge has been raised. There is a stream running into the fort through a very small opening in the wall. The opening, which is only about 5 cm wide, is the only way in.

The gatekeeper is still there, so you explain to him that two of your friends are locked inside the fort. "Oh, dear," says the gatekeeper. "We can't get back in until tomorrow morning. Once I lock up, the gates cannot be opened again until morning. If we try to unlock them from the outside, the alarm will go off. The only way to open the gates without setting off the alarm is from the inside, and it takes six different keys! I am afraid your friends are stuck overnight."

Project-Based Inquiry Science DIG 4

Learning Set 1

The Build a Boat Challenge

5 min.

Students are introduced to the challenge of the Learning Set*—to build a boat from foil that will carry a set of keys for 20 seconds.*

○ Engage

Get students thinking about how boats stay afloat and about the challenge. Make sure students can picture the scenario and appreciate the challenge. You might have students lift a set of keys and think about what kind of boat would be necessary to carry the keys.

TEACHER TALK

"You need to get a set of six keys to your friends who are trapped inside this historic fort so they can get out. You're going to use a boat to do this. The problem is, the boat has to keep the keys afloat for 20 seconds, and all you have to make the boat with is a small piece of foil. Six keys will certainly sink if they drop in the water. What will you do?"

"What about the stream and the opening in the wall?" you ask. "I have an idea. What if we build a little boat and float the keys down the stream and in through the opening in the wall? But, what can we use to build the boat?"

You remember that there are some sandwiches wrapped in aluminum foil still in your backpack. You and your friends decide to build a boat out of foil to see if you can get it to hold the weight of the keys.

"Hold on there!" says the gatekeeper. "If the keys fall into the stream, we won't be able to open the gates at all. You will need to make sure that your boat can float for at least 20 seconds so the keys will make it all the way into the fort."

Your group decides to try. The gatekeeper lets you use the keys and a bucket of water to determine if your plan will work. One problem you have is that the supply of foil is limited, so you can use only a 5" × 5" square of foil. The other problem is that the gatekeeper tells you that he has to leave in 20 minutes!

You begin by discussing the problem with your group, and then you will try to design a boat that can carry six keys downstream for at least 20 seconds. Your group will have 10 minutes to complete the boat.

△ Guide

Show students a 5" × 5" square of foil and a set of keys like the ones they will work with. Explain that groups will have to design a boat that can carry the keys for 20 seconds, using foil and they will only have 10 minutes to design the boat.

SECTION 1.1 INTRODUCTION

1.1 Understand the Challenge

Identify Criteria and Constraints

◀ *$1\frac{1}{2}$ class periods**

*A class period is considered to be one 40 to 50 minute class.

Overview

Students are introduced to the *Big Question: How do scientists work together to solve problems?* and to the *Build a Boat Challenge*, which engages them in the scientific enterprise. They begin to address the challenge by listing the criteria and constraints of the challenge. Groups quickly build a boat and test their initial design ideas. They present their designs to the class, describing the ideas they tried and explaining why they chose their current construction. Finally, they update their lists of criteria and constraints.

Targeted Concepts, Skills, and Nature of Science	Performance Expectations
Scientists often work together and then share their findings. Sharing findings makes new information available and helps scientists refine their ideas and build on others' ideas. When another person's or group's idea is used, credit needs to be given.	Students should be able to describe how their ideas evolved as they were exposed to new information (ideas from other groups or things they read). Students should be able to work effectively together to build a book support. They should listen to each other, try out a variety of ideas, and prepare presentations in which they share their ideas.
Criteria and constraints are important in design.	Students should be able to describe what criteria and constraints are and create a list of criteria and constraints for the *Build a Boat Challenge*.

Materials	
3-4 per group	5" × 5" squares of foil
6 per group	Keys or $\frac{1}{2}$ " hex nut

Materials	
1 per group	Stopwatch Bucket or container of water Paper towels, to clean up spills Scissors Masking tape Stream table
1 per student	*Boat Records* page

Watch out for spills — they should be cleaned up right away to avoid slipping.

Activity Setup and Preparation

Select student groups of three to four students. These groups should stay together throughout this *Learning Set*. There should be at least five groups and no more than eight.

Make a boat yourself before class and test it. See how you can put keys in the boat without tipping it over. Decide if you want to run the tests on the students' designs or if they will.

Cut out 5" × 5" foil squares for each group. You will need about four per group.

Think about how you want to distribute the materials. Each group will need a bucket of water to test their boats. You will need to get these to their workstations with minimal spilling. Groups will also need 5" X 5" squares of foil and sets of keys. You may want to prepare small bags or trays for these materials for each group.

Homework Options

Homework options provide students with short, relevant work options. Each of these assignments encourages reflection on the current section or prepares students to make the connections between one section and the next. A variety of homework option tasks are included in different parts of the Unit.

Reflection

- **Science Process:** Why are criteria and constraints important in design? *(Students should recognize that criteria and constraints are important in design because they describe what is needed and the limitations of the design.)*

- **Science Process:** If you could ignore the constraints of the challenge, what kind of boat would you design? *(Students should think creatively, perhaps drawing on their personal experience with boats. Their answers may be useful later in pointing the way to a solution.)*
- **Science Process:** You are building a treehouse. The kid across the street has a treehouse 10 feet off the ground. You want your treehouse to be at least 10 feet off the ground, but the branches get too thick at about 18 feet off the ground, so the roof of your treehouse cannot be higher than 18 feet. You want to put a stereo in the treehouse, but the electrical cord will only reach the side of the treehouse near your house, so you have to put the entrance there. What are your criteria and what are your constraints? *(Criteria: The treehouse must be 10 feet off the ground; stereo in the treehouse. Constraints: roof of the treehouse must be 18 feet high or less; entrance must face the house.)*

Preparation for 1.2

- **Science Process:** Sketch and explain two or three possible boat designs you would like to try based on what you now know and the updated criteria and constraints. Do not forget to draw a diagram and list the ways your design fulfills the new criteria and constraints. *(Check for students' accuracy of criteria and constraints).*

NOTES

SECTION 1.1 IMPLEMENTATION

1.1 Understand the Challenge

Identify Criteria and Constraints

criteria (singular: criterion): goals that must be satisfied to successfully achieve a challenge.

constraints: factors that limit how you can solve a problem.

Before you start, it is a good idea to make sure you understand what your challenge is. Design challenges have two parts: criteria and constraints.

Criteria are things that must be satisfied to achieve the challenge. For the boat challenge, this will include the job the boat must do and the specific details of how the boat will do that job. To get the job done, you may have thought about buying a boat. However, the challenge does not offer that as a possibility. So, one **criterion** is that you must build the boat.

Constraints are factors that limit how you can solve a problem. Think about the constraints that have been placed on you for this challenge. You can use only a 5" × 5" square of aluminum foil. This makes the size of the foil a constraint. Think about other constraints that have been placed on you for this challenge.

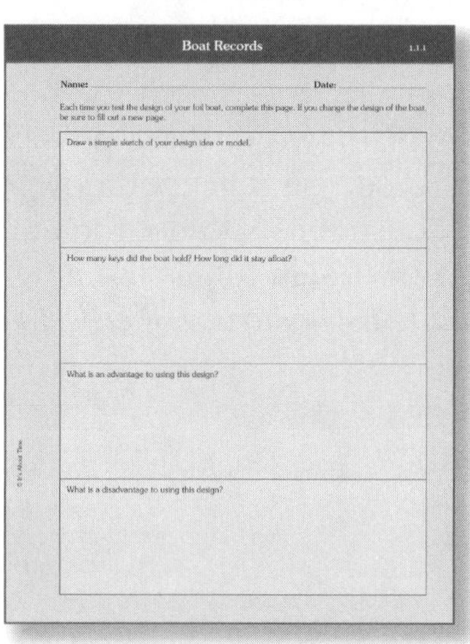

Build and Test Your First Boat

To help you think about how to achieve your challenge, you will begin by drawing a sketch of what your boat will look like. As you sketch your boat, think about how the product you are designing is supposed to function, or work. Remember that you have only a 5" × 5" square of foil to work with, and the boat you build must be able to carry six keys for at least 20 seconds. After everybody has had a chance to think and sketch, spend some time sharing your ideas with your group members. Then work together to build and test your boat. Try out different ideas. Think about which ones seem to work better. You will have a total of 10 minutes.

Project-Based Inquiry Science DIG 6

1.1 Understand the Challenge

Identify Criteria and Constraints

10 min.

Guide the class in identifying the criteria and constraints for the Build a Boat Challenge.

META NOTES

When groups present their boat to the class, they will check if the designs meet the criteria and constraints of the class. After the presentations, the class will update their criteria and constraints.

△ Guide

Ask students what they think are some of the things they need to accomplish in the boat challenge. Ask what some of the limitations of their solutions are.

As students identify criteria and constraints, record them on a class list. When there are several items for each, introduce the terms criteria and constraints. Criteria are goals that must be satisfied and constraints are limitations.

Emphasize that the class will use the criteria and constraints listed to assess their designs and that they will have an opportunity to redesign their boats later.

◇ Evaluate

Make sure the following criteria and constraints are listed before moving on. If students are having difficulties, refer to the *What's the* Big Challenge? section, where criteria and constraints are described.

Criteria: the boat must carry a set of six keys; the boat must stay afloat for 20 seconds.

Constraints: the boat must be built in 10 minutes and the boat must be made from a 5" × 5" piece of foil.

Build and Test Your First Boat

20 min.

Student groups design, build, and test their first boat.

△ Guide

Tell students they are now going to design, build, and test a boat. They will first come up with a quick design and then they will present it to the group. Each student should contribute their thoughts to the group. They should test all their ideas using a bucket of water. The group should decide on their best design and build it. Remind them to think about the list of criteria and constraints.

⬡ Get Going

Assign students to groups, distribute the materials, and tell students they have 10 minutes to design and build their boat. Emphasize that each group member should record sketches of the group's designs on their own *Boat Records* page.

△ Guide

While groups are working, visit them and remind students that everyone should contribute their ideas. You may have to model polite ways to discuss ideas, "I disagree with this part of your design because I think ... let's try it out and then let's try out...." Emphasize that students should contribute their thoughts openly and in a polite and considerate way. Remind them to record their designs on their *Boat Records* pages.

> **META NOTES**
>
> It is O.K. for students to struggle with the design. They will refine their ideas and build a boat two more times.

□ Assess

While visiting groups, take note of what issues students have with designing and constructing their boats, and assess their collaborative skills. Focus on these points later while guiding the class discussion and future group work.

Make sure that you only allow students 10 minutes, as this is one of the constraints of the challenge.

Communicate Your Work

25 min.

Students share their designs and how they meet the class list of criteria and constraints.

META NOTES

In *PBIS* students usually share their work with the class in *Solution Briefings*. The presentation of ideas in this segment is less detailed than a *Solution Briefing*. Students will be introduced to *Solution Briefings* in the next section.

Recording Your Results

Keep track of the designs you attempt for this test boat. Record your sketches on a *Boat Records* page. All group members should keep a record of the final boat design on their own pages.

Communicate Your Work

Share Your Designs

It is time to share your design with your classmates. Groups should take turns presenting their boats to the rest of the class. After each presentation, the teacher will test your design to see if it meets the criteria. There will be time for classmates to ask questions of the presenting group.

As you present your boat, try to answer these questions:

- How is your design constructed?
- Why did you design it the way you did?
- How did the challenge constraints affect the design?
- What things did you think about and try before getting to this design?

As you listen to everybody's reports, make sure you understand the answers to these questions for each report. If you do not think you have heard answers to each question, ask questions (like those above). Be careful to ask your questions respectfully.

After each boat is tested, the class should quickly discuss and agree about how well the design fulfills the challenge's criteria.

Materials
- 5" × 5" square of foil
- 6 keys or ½" hex nut
- bucket of water

Reflect

Answer each of the following questions. Draw and label a sketch of your group's boat, showing what its good qualities are, what problems it has, and what you might want to change.

1. Which criteria did your design meet?
2. What qualities make your boat a good design?
3. What are the problems with your current design?
4. What did you learn from other groups that will help you with another design if you have another chance?

△ Guide

Now that students have tried boat designs with their groups, it is time for them to share their designs with the class. This is the first time students have made presentations. Emphasize that they should present to each other, not to you. Emphasize that other students in the class may be interested in what they say and may ask questions. Presenting groups should make sure their presentations answer the questions in the student text. During their presentations, students should speak loudly and clearly and think about what they want to say before speaking.

⬡ Get Going

Let groups know how much time they will have to present. They should have no more than two minutes for their presentations and two minutes for questions. Give groups about two minutes to prepare their presentations. Make sure you have a bucket or other container of water ready to test each group's boat.

Remind students that if they do not understand a design and how it works, they should ask the presenting group to clarify. Emphasize that during presentations, students in the audience should listen for answers to the questions in the student text. Then have the class begin presentations.

△ Guide Presentations and Discussions

This is an excellent time to begin modeling for students how to ask a question. You might want to discuss this before students start presenting. During the discussion, assist students in their questioning and comments by modeling respectful language such as:

TEACHER TALK

"I agree with... because...

I disagree with ... because...

Could you clarify...

How does that meet the criterion...

What did you see (observe) that made you think the design worked/did not work?"

Model appropriate volume and eye contact for the students.

During the presentations, groups should address the four questions in the student text. Give students in the audience a chance to ask questions first, then guide presenters if needed.

1. Help students point out and describe their design features.
2. Remind students to provide the reasoning behind their design. What made them think to do that? What did they try out that led them to that idea? Listen for comments about how well the design floats and how stable it is.
3. Challenge students if you think any constraints have been violated.
4. Ask what they decided against and their reasons why. This helps students recognize that there is valuable information even in ideas that do not work.

META NOTES

Students may not have experience presenting to each other. You may need to give a lot of support to help them get used to presenting to each other and not simply to you, the teacher. It is therefore important during this first presentation to model discussion techniques and assist students in rewording and redirecting their questions as necessary. Read the tips on discussion techniques in the *Teacher's Resource Guide*.

Reflect

10 min.

Wrap up the class discussion by having students reflect on their designs.

At the end of each presentation, test the group's boat by setting it on the surface of the water in the bucket and then carefully setting the keys in the boat, or have each group run the test themselves for the class. Remember the boat should be timed for 20 seconds.

After each boat test, pause to discuss how the criteria and constraints were or were not met. Assure students the ideas that work and the ideas that do not work are important. One gives you ideas to use, and the other gives you ideas that need to be refined or should not be used.

...ell the design fulfills the ... criteria.

Reflect

Answer each of the following questions. Draw and label a sketch of your group's boat, showing what its good qualities are, what problems it has, and what you might want to change.

1. Which criteria did your design meet?
2. What qualities make your boat a good design?
3. What are the problems with your current design?
4. What did you learn from other groups that will help you with another design if you have another chance?

DIG 7

DIGGING IN

META NOTES

Notice how students are communicating with each other. What are they having trouble communicating? Keep these in mind to help them with in later presentations.

META NOTES

Look for common ideas students have that they may want to include as part of their criteria for the boat design.

△ Guide and Assess

1. Students' answers should be honest and reflect an understanding of the criteria. If the boat sank after two seconds, the design did not meet the time criterion, and this should be noted.
2. Students should select qualities that help the boat meet the criteria.
3. Students should describe ways the boat did not meet the criteria. At this point, they probably do not know enough about the science involved to isolate features of the design that prevent the boat from meeting the criteria.
4. Look for signs that students are paying attention to what worked in other groups' designs. If students are having difficulties you might ask which boat designs they thought were good and why.

Look for common ideas students have that they may want to include as part of their criteria for the boat design.

Update Your Criteria and Constraints

Now that you have tried achieving the challenge, you have found that there is more to think about than you imagined earlier. You may now realize that the criteria and constraints are different than you had first expected. For example, you know that six keys weigh quite a bit. You will soon have a chance to design and build a better boat. Before that, review your list of criteria and constraints. Update the list, making it more accurate. A more accurate list will help you design a better-performing boat.

What's the Point?

You now understand what is required for achieving this challenge better than you did when it was first presented to you. People often try to solve a problem without taking the time to think about it first. If you do not understand a problem well, your solution will not be the best one. In fact, you might fail. Each time you are presented with a new problem, take the time to think. Identify what you have to achieve (criteria). Also, consider what limits you are working under (constraints). You may also find it useful to explore the materials you will be using. Making a first, simple try at a solution may give you a better understanding of the problem you have to solve. With better understanding of the problem, and what is required to solve it, you are more likely to be successful.

Update Your Criteria and Constraints

5 min.

Hold a class discussion to refine the class's criteria and constraints based on what they learned from the presentations and discussions of groups' designs.

△ Guide

Guide students' attention to the criteria and constraints of the challenge by pointing out that they have seen a lot of designs and considered how each has met or has not met the criteria and constraints. Now they should reconsider what the criteria and constraints are.

You may want to point out any design features that everyone used but were not required by the criteria in their list. These design features may suggest another criterion that should be in their list. If the class really liked one group's design, you might suggest they pick a particular feature of it to add to the list of criteria.

Edit the list as students make their suggestions.

Ask students to quickly note any ideas they have for their next design.

NOTES

Assessment Options

Targeted Concepts, Skills, and Nature of Science	How do I know if students got it?
Scientists often work together and then share their findings. Sharing findings makes new information available and helps scientists refine their ideas and build on others' ideas. When another person's or group's idea is used, credit needs to be given.	**ASK:** Give an example from what you did today of how scientific ideas and methods change. **LISTEN:** Students should be able to describe how their ideas evolved as they were exposed to new information (ideas from other groups or things they read). **ASK:** When is another iteration is useful? **LISTEN:** Students should be able to describe how each iteration is done to make use of new information and understanding or to gain more information and understanding. **ASK:** How has working in a group and presenting your group's work to the class affected your design plans? **LISTEN:** Students should be able to discuss how they decided upon and built the final group idea based on all the ideas shared in the group, and how they updated the criteria and constraints based on the class presentations and discussions. Some students may say they have new ideas for the boat based on what they saw and heard during the class presentations.
Criteria and constraints are important in design.	**ASK:** Why is it important to know the criteria and constraints in a design? **LISTEN:** Students should be able to define criteria (requirements that must be met) and constraints (limitations), and they should be able to describe how it is important to understand these when designing.

Teacher Reflection Questions

- What evidence do you have that students understand the distinction between criteria and constraints? What can you do to make this distinction clearer?
- One of the goals in this section is to help students learn how to improve their own designs by sharing ideas with each other and trying again. What difficulties did students have with this? How will you handle it next time you teach this section?
- During the presentations, what did you do to assist in making the discussion student-centered? How were you able to get students to address their questions to other students instead of to you?

NOTES

SECTION 1.2 INTRODUCTION

1.2 Design

Build a Better Boat I

◀ $1\frac{1}{2}$ *class periods*

A class period is considered to be one 40 to 50 minute class.

Overview

Groups design and build a new boat using the list of criteria and constraints they updated after seeing each other's initial designs. By designing and building their boats based on new knowledge, students learn about the importance of iteration and keeping good records. Groups present their designs to the class, engaging in a common social practice among scientists—sharing and refining ideas within their community. Like scientists, students may have used ideas from others in order to reach their goals. Students learn that scientists build their ideas on the ideas of others, and they learn about the importance of crediting the work of others.

Targeted Concepts, Skills, and Nature of Science	Performance Expectations
Scientists often work together and then share their findings. Sharing findings makes new information available and helps scientists refine their ideas and build on others' ideas. When another person's or group's idea is used, credit needs to be given.	Students should be able to describe how their ideas evolved as they were exposed to new information (ideas from other groups or things they read). Students should be able to work effectively together to build a boat. They should listen to each other, try out a variety of ideas, and prepare presentations in which they share their ideas.
Criteria and constraints are important in design.	Students should be able to describe what criteria and constraints are, and they should be able to create a list of criteria and constraints for the *Build a Boat Challenge*.
Scientists must keep clear, accurate, and descriptive records of what they do so they can share their work with others and consider what they did, why they did it, and what they want to do next.	Students should be able to keep descriptive and accurate records of designing and building their boat.

Materials	
3-4 per group	5" × 5" squares of foil
6 per group	Keys or $\frac{1}{2}$" hex nuts
1 per group	Stopwatch Bucket or container of water Paper towels to clean up spills Scissors Masking tape Stream table
1 per student	*Boat Records* page *Solution-Briefing Notes* page

Activity Setup and Preparation

Cut out 5" × 5" foil squares for each group. You will need about four per group. Each group will need a bucket of water.

Watch out for spills — they should be cleaned up right away to avoid slipping.

Homework Options

Reflection

- **Nature of Science:** What is the difference between copying and building on someone's idea? Describe examples of copying versus building on others' ideas. *(Students should recognize that copying does not credit another's work; building on someone's idea does. Also, building includes adding or modifying, not just using.)*
- **Science Process:** If you could redesign your boat, what ideas from the *Solution Briefing* would you like to build upon? Why? *(Check that students support their ideas with reasons related to criteria or constraints.)*
- **Science Content:** Based on the ideas from the *Solution Briefing*, sketch a design for a boat. Describe or explain. *(Check that the criteria and constraints are met in the sketch.)*

Preparation for 1.3

- **Science Content:** What makes some boats float and others with the same number of keys sink? How could you make your boat more able to float while carrying a load? Support all your answers with reasons. *(This gets students thinking about the topics that will be covered in the next section. Look for what ideas students have about how the shape of the boat affects how the water supports the boat.)*

NOTES

NOTES

SECTION 1.2 IMPLEMENTATION

1.2 Design

Build a Better Boat I

You have just completed your first attempt at building an aluminum-foil boat. You also talked about the design ideas and products every group created. Along the way, you learned about some new ideas that work well, as well as some ideas that do not work so well. Many of you might like to have another try at building a better boat. You are going to have that chance. In this science class, you will have many chances to modify solutions to problems or challenges. Each try you make is called an **iteration**. Before you start your next iteration of the Boat Challenge, read the rest of this page and the next page to prepare for what's ahead. You will then have 10 minutes to design and build a better boat.

iteration: a repetition that attempts to improve on a process or product.

Plan Your Boat Design

You built your first boat quickly and without a lot of planning. It may or may not have worked well. During this second attempt, you have a chance to design and build a boat that really works. Consider what you learned from your first attempt. Did it meet the criteria of the challenge? If not, what can you do to improve the design so it does meet the criteria? You may get ideas by thinking about the different designs that your classmates came up with. Think about the designs that worked well and those that did not. Discuss these ideas with your group members. This will make your design better.

Be a Scientist

When people design things, they usually call it a product. Often, designers do not create the best or most successful product the first time. Just like you did with your group, they try something. Then they figure out the strengths (what was good), and the weaknesses (what was not good) in the design. They might decide that they need different materials. They might decide that they need to put things together differently. They might decide to make small changes or to make big changes. After the first time, they understand the challenge better. After the second time, they may also find that their solution is not as good as they would like. Designers often try again and again before they get the product just the way they want it. Each try is called an iteration.

DIG 9 DIGGING IN

◄ $1\frac{1}{2}$ *class periods**

1.2 Design

Build a Better Boat I

5 min.

Students revise their designs based on the new criteria and constraints, and ideas from the presentations.

○ Engage

Remind students of the designs they saw and the new ideas they might want to try out based on what they saw and heard from other groups. Emphasize that they will get to rework solutions to problems or challenges a number of times in this course. This time, they will keep records of what they are doing as they redesign their boats.

*A class period is considered to be one 40 to 50 minute class.

TEACHER TALK

"After you saw all the things different groups tried for their boats and you saw what happened when we tried them out, you probably had some new ideas for your boat. Today you'll have a chance to create a new boat. You'll need to keep good records as you try out different things and see what works and what doesn't work."

Plan Your Boat Design

5 min.

Groups record their plans for their revised boat design.

10 minutes to design and build a better boat.

Plan Your Boat Design

You built your first boat quickly and without a lot of planning. It may or may not have worked well. During this second attempt, you have a chance to design and build a boat that really works. Consider what you learned from your first attempt. Did it meet the criteria of the challenge? If not, what can you do to improve the design so it does meet the criteria? You may get ideas by thinking about the different designs that your classmates came up with. Think about the designs that worked well and those that did not. Discuss these ideas with your group members. This will make your design better.

△ Guide

Remind students when planning their designs they should aim to make a boat that satisfies all the criteria and constraints. Tell them to think about the different designs their classmates came up with and emphasize that they should try to build on these ideas.

⬡ Get Going

Each member of a group should take turns presenting their ideas to the group. The group should discuss each idea and consider the strengths and weaknesses of each idea.

Tell groups they have five minutes to share and plan a design and they should record their plans.

△ Guide

While students are sharing their ideas, stop by a few groups and ask them how their designs meet the criteria and constraints, and how they are better than their last design.

Build and Test Your Design

Build and test a working boat that can carry six keys for at least 20 seconds. Keep records of each iteration.

Recording Your Work

Materials
- 5"× 5" square of foil
- 6 keys or ½" hex nut
- bucket of water

Keep track of the number of keys your boat can keep afloat and the amount of time it can stay afloat. Record this information on new *Boat Records* pages. Each of you should fill in your own page.

One part of the *Boat Records* page has room for you to draw your design after you have finished building it. Each time you create a new design, continue to fill out the page. This way you will have a record of all the designs you attempted. You will be able to see how you made improvements along the way.

After completing your boat, you will present it to the rest of the class. During your presentation, have your *Boat Records* pages handy to report your results to the class.

The designers of this boat went through many iterations to improve the boat's design.

Build and Test Your Design

15 min.

Groups record each iteration as they build and test their boats.

△ Guide

Focus students on testing, revising, and retesting their designs. Point out the importance of reworking a solution to a problem or challenge and that multiple attempts, or iterations, will help them improve their designs. Point out that designers and scientists go through many iterations in their work. Emphasize that an iteration is not just redoing the same thing. It is first modifying the plan and then trying the revised plan.

TEACHER TALK

“All of you have just redesigned your boat. You have tried to make it better based on all the ideas you heard during the class presentations. Designers and scientists rework solutions to problems or challenges to improve them. Each reworking is called an iteration. An iteration is a modification and retesting in design used to make something better. Your current design plan is your second iteration.”

Focus students' attention on the importance of record keeping. Point out that multiple iterations of designing and testing will require them to keep records to be able to keep track of the design changes and remember how each version worked. Emphasize the four points in the *Be a Scientist: Keeping Records* box.

TEACHER TALK

“Each iteration is an improvement of your previous design. You might forget what you have done if you don't keep good records. Keeping good records is important. It helps you share your work with others. It helps you to remember what you did, why you decided to do it, how you did it, and what you think you should do next. When scientists keep records they want them to be very descriptive and provide their reasoning. Scientist write down what went right, what went wrong and what not to do next time.”

⬡ Get Going

Have student use the *Boat Records* pages to record their designs using one page per design. Tell each student to write their reasoning for their design in the first row.

TEACHER TALK

“The *Boat Records* page has three parts to it. In the first part you will draw your design and describe your reasons for choosing that design. In the next space, you will record how many keys your boat could hold and how long it could stay afloat. In the last spaces you are going to evaluate your design. You will record the advantages of the design and then the disadvantages of the design. If you made your boat very wide so it had a lot of support on the bottom, but water leaked in over the sides there is an advantage and a disadvantage to that choice. The advantage is the support you get on the bottom and the disadvantage is that water leaks in on the sides. You can create additional designs while there is time. You should try out the group's favorite design ideas first.”

Inform groups they have 10 minutes to construct their designs, test each one, and then select the best design to present to the class. Distribute the materials or have students pick up the materials at a materials station.

△ Guide

While groups are building their designs, assist them in filling out the *Boat Records* page if they are having difficulty. They may need to be reminded to write down why they chose that particular design. To help them identify why they chose a design, you could ask what did not work in their last design, what they could improve and what they want to retest.

If you notice some groups are having trouble collaborating, assist them with their group skills by encouraging each member to take a minute to present their ideas to their group and to decide which designs they want to try out first.

Let students know when it is time to start wrapping up their testing and to select their best design. This will be when they have about five minutes left.

◇ Evaluate

Check to see how groups are progressing. If most groups are not done in 10 minutes, give them another minute. If most groups have tried out a few of their design ideas and have decided on their best one, then transition groups to preparing to present their results.

NOTES

Communicate Your Results

Solution Briefing

20 min.

Students prepare for their first Solution Briefing.

Be a Scientist

Keeping Records

Scientists always record their work as they go. To record means to write, illustrate, or diagram what is being done. This important step allows scientists to accurately report their findings to others and helps them design future investigations.

You probably had to keep records of your work in science class before. Keeping records is important for scientists, and it is also important for you as a student scientist. Recording your work helps you to do the following:

- Share your work with others.
- Remember what you did and decided along the way.
- Remember why you decided to do those things.
- Make decisions about what investigations to do next.

Communicate Your Results

Solution Briefing

After you designed and built your first boat, you presented it to the class. You will also present your redesigned boat to the class. This time you will present more formally in a *Solution Briefing*. In a *Solution Briefing*, you present your solution in a way that will allow others to evaluate how well it achieves criteria and to make suggestions about how you might improve it. Before you start preparing, read more about *Solution Briefings* on page 13.

Get ready for this briefing by preparing a presentation that answers these questions:

- How is your design constructed?
- What materials did you use?
- Why did you design and build it the way you did?
- How does the design meet the criteria?
- How did the constraints affect the design?

△ Guide

Describe to students what a *Solution Briefing* is and how it works. Explain that they will be presenting their works in progress to share their ideas and gather advice. The goal is for the larger group (the class) to help each small group to make their solution better.

TEACHER TALK

"Now that you have all worked for a while on improving your designs, it is time for you to get some advice from outside of your group. Designers (and scientists) often get together to share their ideas and get advice from each other on ways to improve their designs (or explanations). *PBIS* calls this a *Solution Briefing*. We are going to have our first *Solution Briefing* so you can share your boat designs and get advice from the class. Everyone is encouraged to ask questions so they understand your design and then to offer suggestions on ways to improve it."

Describe how to prepare for a *Solution Briefing*. Explain that groups' presentations should describe how they arrived at their current design solution. They can use the questions in the student text to help think about what to present. Explain that the audience should ask clarifying questions and offer suggestions. Everyone should voice their questions and ideas in a polite and considerate manner, using language such as "I agree with ... because..." or "I disagree with... because."

META NOTES

Remember this is the first time students have seen a *Solution Briefing*, and it will take time for them to understand how to participate as presenters and as the audience.

TEACHER TALK

"It is important that you think about what you are going to say when you present your designs. In your books, there is a list of questions that will help you think about what to talk about during your presentation. When you are in the audience or not presenting you will need to listen to make sure that you understand each design. You should also voice your opinions and ideas. Remember to always be polite and considerate whether you are presenting or are in the audience."

⬡ Get Going

Give students five minutes to prepare their presentations.

Lead student groups in the *Solution Briefing*. Point out the *Solution-Briefing Notes* page for students to keep track of presented designs and ideas they found useful. Remind students that each group will be presenting their best design and that the audience will be asking questions and giving advice.

Remind the class to listen for how the designs work and to prepare their questions for the presenting group.

- What past experiences helped you make your design?
- What problems remain?
- What things did you try along the way?
- How well does your boat work? What else do you want to test?

As you listen to your classmates' presentations, make sure you understand the answers to the questions on the previous page. If you do not understand something, or if they did not present something clearly enough, ask questions. You can use the questions on the previous page as a guide. When you think something can be improved, make sure to contribute your ideas. Be careful to ask your questions and make your suggestions respectfully. Record the interesting things you are hearing on your *Solution-Briefing Notes* page.

Solution-Briefing Notes

Name: ______ Date: ______

Design Iteration: ______

Design or group	How well it works	What I learned and useful ideas		
		Design ideas	Construction ideas	Science ideas

Plans for our next iteration

△ Guide

While groups are presenting, you and the class should be listening for descriptions of each design, including how it is constructed and how well it works to carry a set of six keys.

If groups get off track or stuck, you can prompt them with one or two of the nine questions listed in the student text. As the *Solution Briefing* progresses, encourage students in the audience to participate in asking questions to help the presenters clarify their design descriptions.

Be a Scientist

Introducing a *Solution Briefing*

A *Solution Briefing* is useful when you have made one or more attempts to solve a problem or achieve a challenge and need some advice. It gives you a chance to share what you have tried and learned. It also provides an opportunity for you to learn from others. You can ask advice of others about difficulties you are having.

Real-life designers present their designs to each other and to others several times as they work on design projects. A team of designers sets up their design or design plans, and everybody gathers around. They make sure everyone can see. The design team presents their design plan to everyone. The other designers ask questions and give helpful advice about ways to improve the design.

You will do the same thing. In a *Solution Briefing*, each team presents their solution for others to see. Then teams take turns presenting to the class. Other classmates ask questions and offer helpful advice. You might walk around the class from design to design, or teams might take turns presenting in front of the class.

A *Solution Briefing* works best when everyone communicates well. Before you present your design to the rest of the class, think about what might be important to share. What aspects of your design should you present? What parts do you want to discuss with others? You need to be ready to justify to others what you decided to do and why.

When you are listening to a *Solution Briefing*, it is important to pay close attention. Look at each design or plan. Think about questions you would like to ask about the design.

Each time you hold a briefing, you will take notes. You will fill out a *Solution-Briefing Notes* page as you listen to each group's presentation.

Collaboration is a group effort. A team of designers share sketches and ideas at the innovation and design firm, IDEO.

Use the following guidelines to assess how students respond to the questions in *Communicate Your Results.*

- *Solution Briefings* should give detailed descriptions of groups' designs, pointing out important design features. You could ask students what features of their designs have changed or which parts of their designs they think are most important.
- For this challenge, everyone used foil, as this is one of the constraints.

- *Solution Briefings* should give reasons for the features of the design, explaining how the criteria and constraints of the challenge guided their design choices. You could ask why they chose each part of the design and why they chose to make the changes in their last design.
- Students should explain how their designs met each of the criteria. Students may provide this explanation with the reasons for the features of their designs.
- Students should describe how each constraint affected their design. They may describe the reasons for the features of their designs.
- Presentations should recount past experiences (including the previous class period) and how they used those experiences to make design choices. You could ask if there were any designs from the previous class that helped them decide on this design.
- Presentations should identify problems that students feel remain and should suggest ideas they think might be good solutions to these problems. You could ask what they think could improve. Some students might mention features in other groups' designs that they think are effective.
- *Solution Briefings* should include histories of the ideas groups tried. You could ask what else they tried or thought about trying. It is important to recognize what ideas did not work, as well as what ideas worked. Point out that thinking about the ideas that did not work is important for developing a complete understanding of what works.
- Students should include an assessment of how well their boats worked. They should also discuss some ideas they would still like to test.

After all groups have presented, wrap up the *Solution Briefing* by asking students if they heard any interesting ideas or thought of new ideas they would like to use to improve their boat design. Guide students to think about how sharing ideas will help them improve their own ideas.

Compare this session with what scientists and designers do to improve their solutions. Reflect on the usefulness of iterations and keeping good records. Tell students that scientists are always building on each other's ideas. Bring up the idea of giving credit to others when you build from their ideas. This is not copying. Emphasize that the difference between this and copying is that copying does not give credit to the people who thought of the ideas you used.

META NOTES

If students say that another group copied their idea, tell them this is an important issue and will be addressed in a little bit and continue with the presentations. After the presentations the class will discuss building on each other's ideas versus copying.

Reflect

Following your *Solution Briefing*, answer the following questions. Discuss your answers and how they may help you better achieve the *Boat Challenge*.

You may find looking at your drawings and your *Solution-Briefing Notes* page helpful as you answer the following questions. Be prepared to discuss your answers with your class.

1. Before each boat design was tested, what did you think would happen when the keys were placed in it?
2. Which boats worked the way you thought they would? Which worked differently than you expected? For the ones that worked differently, what might help you to understand why?
3. What qualities make your boat a good design?
4. What are the problems with your current design?
5. What can you borrow from other designs to make yours work better?
6. What do you need to learn more about to make a better design?

Be a Scientist

Build on and Benefit from One Another's Ideas

You ask questions and offer suggestions during a *Solution Briefing*. When you do this, you are **collaborating** with one another. You are working together. You offer your ideas for others to think about. You provide suggestions that might help them improve their solutions. Sometimes you learn something that you want to try yourself.

Other teams may come up with solutions or ideas that you want to borrow and make better. You may also find that other teams have used your suggestions. Is the other team copying from you? Are you copying from them?

Think about the other team's success as your success when they use something you suggested. Help them see that your success is theirs if you borrowed something from them.

collaborate: to work together.

Reflect

15 min.

Groups reflect on their designs and create plans for their next iteration based on what they now know.

META NOTES

Groups will build their next boat in *Section 1.4* and they will have to meet a new criterion: the boat must carry eight keys, not six. Do not introduce the new criterion until *Section 1.4.*

⬡ Get Going

Transition the class by reminding them they have just been introduced to many different ideas on how to build the boat. By this time, they should have ideas about how to improve their boat. The six *Reflect* questions will assist students in achieving the boat challenge. Groups should work together at coming up with their best group answer to each question.

Tell groups they have five minutes to answer the questions.

△ Guide and Assess a Discussion

Hold a class discussion. Once groups have had a chance to answer the questions, focusing on Questions 2 and 6. You may want to ask a group or two to answer Questions 1, 3, 4, and 5, but you should get each group to discuss their responses to Questions 2 and 6. These questions help students to create a better design.

Listen for the following:

1. Students should discuss what they expected to happen with specific designs, focusing on any designs that had distinctive or interesting features.
2. This question will help prepare students for creating their next design iteration. Groups should describe the design features that worked as they expected and those that did not. Ask students why they think some things did not work. Why do they expect their new ideas to work? Note which designs surprised students in the ways they worked. You can revisit these designs when the class has learned the science knowledge to explain them.
3. Students should discuss qualities that enable the boat to meet the design criteria, and they should back up their claims with reasons.
4. Students should support their responses with reasons, such as it does not meet the criteria and/or the constraints. Students may also pick out a problem after deciding they liked a feature of another group's design better.
5. Students should focus on features that enabled a design to meet the criteria.
6. If students struggle with this question, remind them of the things that surprised them. Point out different features of designs that students examined during presentations and that students might need to know more about to make good design choices.

⬡ Get Going

Now that students have thought about their designs and what worked and what they would do differently, they will have new ideas about designing their boats. This is the time to complete the S*olution-Briefing Notes* page by filling out the next iteration of their design.

Ask groups to complete their *Solution-Briefing Notes* page based on what they now know. Let them know they may work together as a group but each student is responsible for completing the plans for the next iteration. Emphasize that they should give their reasoning behind their design. Let them know they have five minutes to complete this.

When movie actors receive awards, they often thank many people. Even though they are getting the award, they know that it takes lots of people to put a movie together. It is the same with scientists and engineers.

They would not be able to solve problems or learn new things without building on the work of others. Scientists and engineers write papers, or articles, in journals. They tell others what they have discovered. Others read those papers, talk about the ideas, and ask questions. When someone improves on an idea, they write a paper about it and publish it for others to read. Other people can then improve on their ideas. Each time someone publishes a paper, they give credit to the scientists or engineers they got their ideas from. They do this by **citing** the paper that the idea was presented in. This is the way science is done.

Engineers and inventors submit patents, telling others how to make something work. When other engineers and product designers use these ideas in their own products, they give credit and pay **royalties** for using those ideas. This is how product design and invention are done.

cite: to quote something or somebody.

royalty: in this case, a fee paid to the owner of a patent for permission to use it.

These students, like a group of scientists, are collaborating on their design project. They are all contributing questions and information. This sharing of ideas and building on the past work of other scientists are the ways the best science research is accomplished.

△ Guide

As groups are working, check their progress and guide them by asking questions that help them articulate the reasons behind their design. For example, "Why did you shape the bottom of your boat that way?"

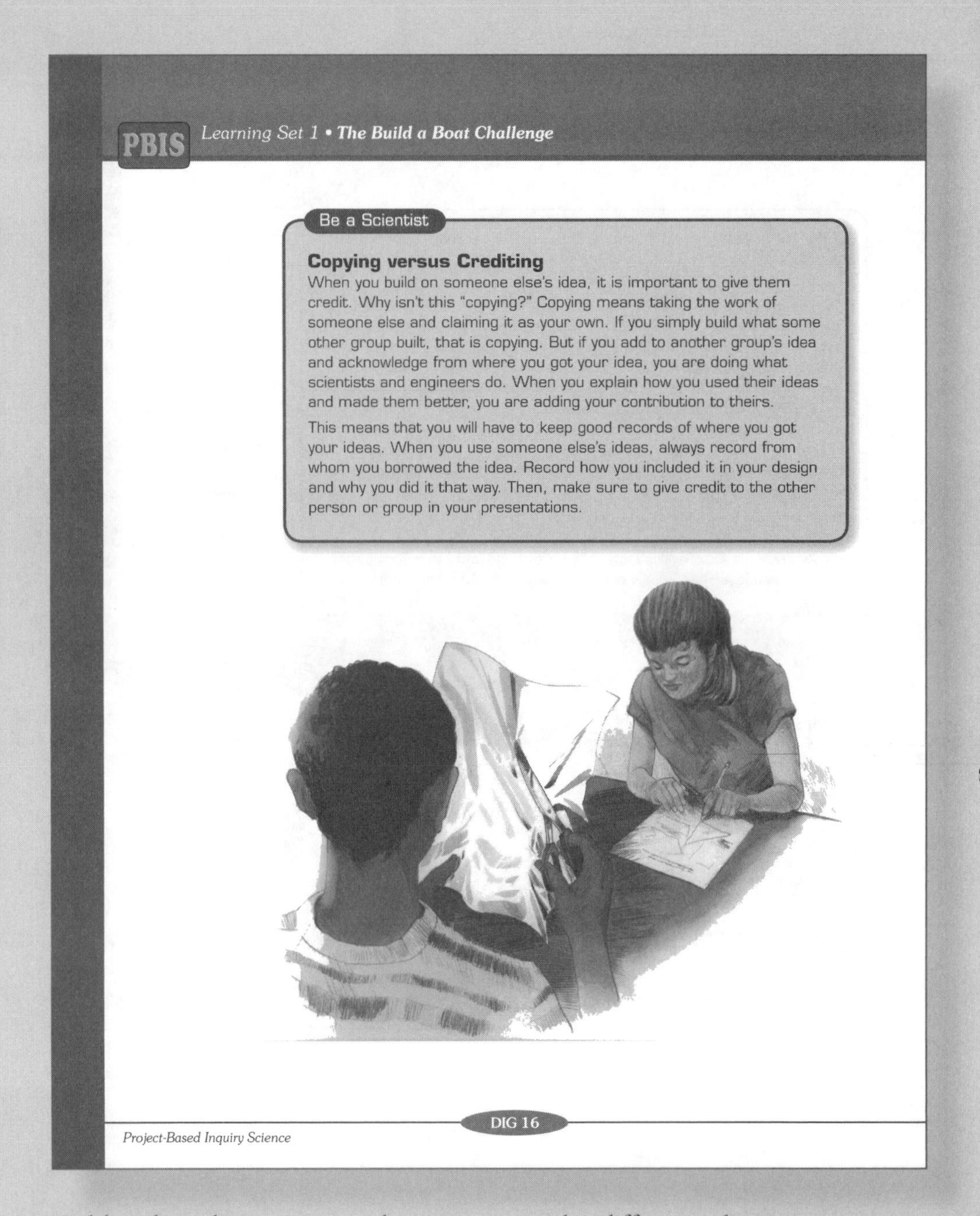

Be a Scientist

Copying versus Crediting

When you build on someone else's idea, it is important to give them credit. Why isn't this "copying?" Copying means taking the work of someone else and claiming it as your own. If you simply build what some other group built, that is copying. But if you add to another group's idea and acknowledge from where you got your idea, you are doing what scientists and engineers do. When you explain how you used their ideas and made them better, you are adding your contribution to theirs.

This means that you will have to keep good records of where you got your ideas. When you use someone else's ideas, always record from whom you borrowed the idea. Record how you included it in your design and why you did it that way. Then, make sure to give credit to the other person or group in your presentations.

Hold a class discussion emphasizing again the difference between copying and giving credit. Emphasize that it is O.K. to build on others' ideas as long as credit is given. There are good examples given in the textboxes of the student text. You may want to go through these. Make sure you introduce the word collaborate and let students know that it means to work together, as they have been doing.

What's the Point?

The boats you built the second time were probably more successful than the first ones. In general, the more chances you get to iterate, the better your solution will turn out. Each attempt you make is an iteration. Each time you make another attempt, you can do better because you use knowledge gained from the previous attempt. You also identify more that you need to learn each time you try to achieve a challenge. Answering the questions that come up before trying again gives you a chance to do even better. This iterative approach to design and problem solving is what scientists do. Use this approach whenever you have a problem to solve.

You read that there is a difference between copying and building on the ideas of others. As you took part in the first *Solution Briefing*, you may have seen some design ideas that worked well. You may have used some of those ideas to improve your boat. Others may have used some of your ideas. When you claim someone else's ideas as your own, it is copying. If, however, you give credit to a group for their idea, you are building on the work of others. This is how scientists work and how science grows. Science builds on the ideas of others.

You probably have begun to realize the importance of keeping records as you work on a design challenge. You used your records when you presented your ideas during the *Solution Briefing*. You also got a chance to see other solution ideas during the briefing. You saw what works and what does not work as well. This may help you develop better ideas as you continue to try to achieve your challenge. You can learn a lot from attempts that "failed" as well as ones that succeeded. In either case, the goal is to understand the challenge better and create better solutions.

Assessment Options

Targeted Concepts, Skills, and Nature of Science	How do I know if students got it?
Scientists often work together and then share their findings. Sharing findings makes new information available and helps scientists refine their ideas and build on others' ideas. When another person's or group's idea is used, credit needs to be given.	**ASK:** Did your group change their design plans based on ideas and input from more than one group member? Did your group get ideas from one or more of the other presenting groups? How many different people contributed to our knowledge base of building a good boat? What should we do when we build on someone's ideas? **LISTEN:** Students' responses should refer to the need to give credit. This is an ethical issue and is important in the scientific community. **ASK:** Do you think our knowledge may be considered a constraint? Why? **LISTEN:** Students will have different answers. This is to prepare them for the next section, but most students should realize that our knowledge is limited, and as we learn more we can create better designs. **ASK:** In what ways do you think we did what scientists do? **LISTEN:** Students should explain how scientists first work in small groups and then share their ideas with a larger community to make sure their ideas are as good as they can be.

NOTES

Targeted Concepts, Skills, and Nature of Science	How do I know if students got it?
Scientists must keep clear, accurate, and descriptive records of what they do so they can share their work with others and consider what they did, why they did it, and what they want to do next.	**ASK:** Why is it important to keep good records? **LISTEN:** Students' answers should contain information about being able to accurately tell others of their work, what they did, and why they did it. **NOTE:** If students don't explicitly make the connection between the records and sharing of results with the *Boat Record* page and with the *Solution Briefing*, then explicitly make the connection for them.

Teacher Reflection Questions

- What difficulties did students have in understanding the importance of record keeping, sharing ideas, and building on others' ideas by giving credit? Where will your students need the most help with these ideas in the next lesson?
- What difficulties did students have in participating or understanding the social practices of working in small groups and holding presentations and discussions in larger communities? What might be help students participate in the next small group or whole class sharing?
- What questions or comments did students ask or say that might be rude or inappropriate? (For example, "S/He copied from us!" Or talking out of turn.) What will help your students to learn how to question or comment respectfully?

NOTES

NOTES

SECTION 1.3 INTRODUCTION

1.3 Read

The Science of Boat Design

◀ ***1 class period***

A class period is considered to be one 40 to 50 minute class.

Overview

Now that groups have completed two iterations of designing and building a boat, they are introduced to the concepts of matter, gravity, buoyancy, density, and what makes a boat float. Students collect and graph the class's data. After reflecting upon the reading, students consider how they could apply these concepts to their boat design.

Targeted Concepts, Skills, and Nature of Science	Performance Expectations
Science and engineering are dynamic processes, changing as new information becomes available.	Students should describe how their ideas evolved as they were exposed to information involving the science behind their boat design. Students should describe how each iteration is done to make use of new information and understanding or to gain more information and understanding.
Earth's gravity pulls things toward Earth.	Students should describe gravity as a pull between two objects, and why the boat sinks if the push up from the buoyant force is less than the pull down by Earth.
When an object is immersed in a liquid, it experiences an upward force on it called the buoyant force (which is equal to the weight of the amount of fluid the object has displaced).	Students should describe how buoyant force is caused by the water particles pushing up on the object, and how it increases as the surface area of the object increases.
Density is the mass per volume of an object.	Students should describe and distinguish mass, volume, and density using examples such as a box of books and a box of bubble wrap.
The density and surface area of an object determine if the object will sink or float.	Students should describe how an object less dense than the fluid it is placed in will float, and how increasing the surface area of the object increases the buoyant force.

Materials	
2 per student	5" × 5" squares of foil
1 per class (Optional, for demonstration)	Two identical boxes, one filled with books, one filled with bubble wrap Jar filled with oil and water Bucket of water 5"x 5" square of foil, balled-up 5"x 5" square of foil, flat

Activity Setup and Preparation

Cut out two 5" × 5" squares of foil for yourself and two for each student. You will use these to demonstrate how a crumpled piece of foil and a flat piece of foil will rest on a surface. You will also demonstrate what happens when you put them in water. These demonstrations are to be done at different times during the class.

Homework Options

Reflection

- **Science Process:** How are the boat's structures you observed during the class presentations similar to each other? How are they different? *(Students should describe how various designs are similar and different. Similarities should include satisfying the criteria and constraints. Students may point out the width or buoyancy of the design.)*

Reflection and Preparation for 1.4

- **Science Content:** You want to use an inflatable raft to cross a river. You get tired blowing it up, and decide that blowing it up halfway is good enough. As soon as you get on it, it sinks. When you blow it up the rest of the way, it floats—even after you get on it. Explain why. *(Look for answers that include ideas of how density and surface area affect the buoyancy of an object. When the raft is fully inflated, the raft and its cargo (you) are less dense than the water—when it is half inflated the raft plus cargo are not less dense than the water. The raft has a larger surface area in the water and the fully-inflated raft has a larger surface area in the water because it does not bend or collapse when you sit on it.)*

NOTES

*1 class period**

1.3 Read

The Science of Boat Design

5 min.

Students' read about how science ideas relate to how boats float.

SECTION 1.3 IMPLEMENTATION

1.3 Read

The Science of Boat Design

matter: anything that has mass and takes up space.

density: the amount of matter in a given amount of space.

buoyant force: the upward push that keeps objects floating in liquid.

volume: the amount of space that something takes up.

atom: a small particle of matter.

molecule: the combination of two or more atoms.

You have just finished your first attempts at building an aluminum foil boat. You also talked about the design ideas and products of other groups. You discovered some ideas that worked well and others that did not. You identified some questions that you want answered before you try again. Soon you will have another chance to build a better boat. Before you do, you will read and think about the science concepts that explain how boats work, and you can then apply this knowledge to your next boat design. To understand what makes things float, it is important to learn about three science concepts—**matter**, **density**, and **buoyant force**. They are all important to making your foil boat carry more keys.

Matter

All objects of any form (solid, liquid, or gas) are made up of matter. All matter has mass and takes up space. The amount of space that something takes up is its **volume**. The boat you are trying to build is made up of matter, and so is the water the boat floats on. Matter is made of extremely small particles called **atoms**. These atoms combine with other atoms to form larger particles called **molecules**. Molecules attach to one another to form all the objects that you see, touch, hear, taste, and smell.

Density

One factor that affects whether or not something can float is its density. Density is the scientific word for the amount of matter in a certain amount of space. It is a measure of how tightly the molecules making up matter are packed together in the space. The more room the molecules have in a given space, the less dense the matter will be.

If you have a cardboard box full of plastic bubble wrap, it will be lighter than the same-sized box full of books. The different materials in each box take up the same amount of space, but each contains a different amount of matter. Since a book-filled box has more matter than a box filled with bubble wrap, the box of books has greater density.

DIG 18

Project-Based Inquiry Science

○ Engage

Let students know that they will be doing a revised boat challenge and it will be helpful to learn some science ideas related to structures. By researching what is known—in this case, what is known about boat science—they will be able to design a better boat.

*A class period is considered to be one 40 to 50 minute class.

TEACHER TALK

"You've already completed two iterations of your boat design. You've learned from each other's designs and heard each other's advice. When you are designing something, it is also important to find out what research has already been done that pertains to your design. In building your boat, it is useful to know about the science of boats. That is what we will be doing today—learning about the science of boats. This will help you to build a better boat."

△ Guide

Matter

5 min.

Introduce the idea of matter. Lead students in thinking about each of the ideas in the reading. Highlight that matter is anything that has mass and takes up space, whether it is a solid, a liquid, or a gas. The boat, the water, Earth, and the air are all examples of matter. Many students confuse volume and mass. Assist students in understanding the difference between volume and mass. Volume measures how much space something takes up, and mass is how much matter (or material) it has. If you have a box of books and a box of bubble wrap, show these to the students and pass them around as you discuss this.

TEACHER TALK

"Although both volume and mass measure how much "stuff" something has, they are very different. Volume measures how much space something takes up and mass measures how much matter it has. Imagine I had two identical boxes. I could fill one up with bubble wrap and one with books. Both take up the same volume, but the box of books has more mass than the box of bubble wrap."

NOTES

Point out the connection with the challenge. Both boats and water are made of matter.

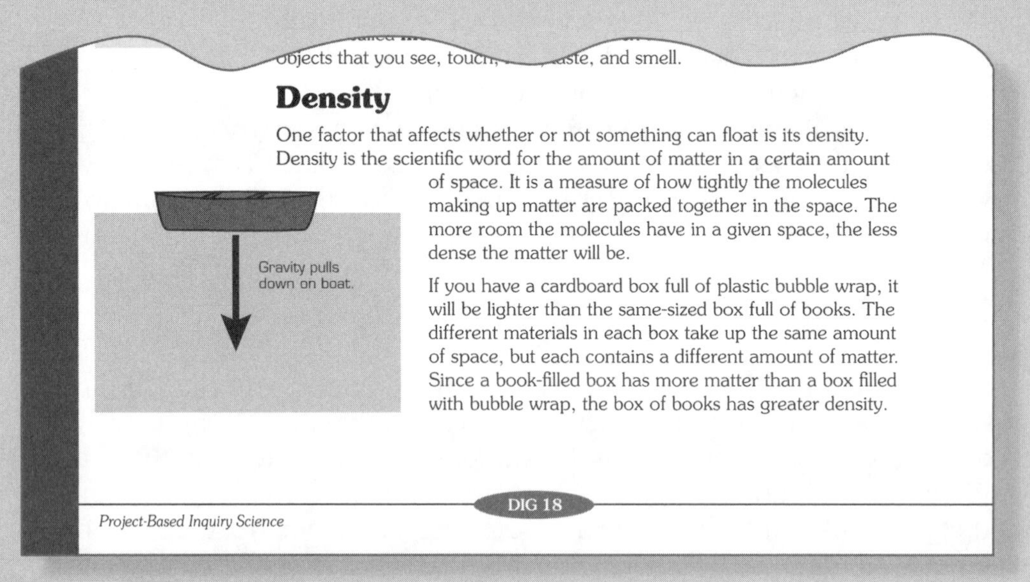

Density

5 min.

Introduce density to the students, which also affects the buoyant force.

META NOTES

The ideas of volume, mass, and gravity are mentioned here to support the ideas of density and buoyancy of boats. Do not spend a lot of time exploring these ideas beyond what is needed for boats. These concepts will be explored in another Unit.

△ Guide

Describe density as the amount of matter in a certain space. Lead students to connect density to the concepts of matter and volume, and make sure they can distinguish density, matter, and volume.

TEACHER TALK

"Think about the example of a box filled with bubble wrap and a box filled with books. Both boxes take up the same amount of space, but the one with books has more matter than the one with bubble wrap.

Which has more mass, a gram of bubble wrap or a gram of books? They're actually the same, right? A gram is a gram. The difference is that the gram of bubble wrap takes up more space.

We say that books are denser than bubble wrap because there is more book mass in a certain volume than there is bubble wrap mass in the same volume. There is more mass in a box of books than in a box of bubble wrap. If you wanted to have a box of bubble wrap with the same mass as a box of books, the box of bubble wrap would have to be much larger than the box of books.

Density describes this characteristic. It describes how much mass there is in a given volume. Mass describes how much matter there is and volume describes how much space it takes up."

Connect the idea of density to boat design. A boat that is denser is more likely to sink than a boat that is less dense.

TEACHER TALK

"Imagine tossing a box of bubble wrap and a box of books into a lake. Which do you think would float and which do you think would sink? The denser one would sink, right? Why?"

NOTES

META NOTES

Some students may believe that air has no mass or that it takes away mass, causing things to float or become less heavy. They might think that putting the box of books on top of the box of bubble wrap somehow makes the books lighter. Although the box of books and the box of bubble wrap taped together might have an overall density and surface area that would allow water to support them, this is not the case. If this comes up, emphasize that air, like all gases, is made up of atoms which have mass. Some gasses are less dense than others, so they rise due to buoyancy. A helium balloon in air rises because helium gas is less dense than air. A cork pushed into water will rise because it is less dense than water.

META NOTES

Consider demonstrating the difference between liquids of different density using a jar with oil and water in it. Point out that the oil is denser than the water and so it sinks to the bottom of the jar.

Buoyant Force

10 min.

We think of books as being heavier than bubble wrap, but that is misleading. What we are really thinking about in that case is the density of the materials. A book taking up a certain amount of space will be heavier than a piece of bubble wrap taking up the same amount of space. This means that books are denser than bubble wrap. There is more matter in a book than in a section of bubble wrap of the same size as the book. That is why a box of books will be heavier than the same box filled with bubble wrap. For the same volume, the more dense material will be heavier than the less dense material.

Molecules that make up the matter in a book are tightly packed together and do not have much space between them. In the bubble wrap, the molecules have a lot of space between them because each bubble contains a lot of air—air is a gas and is much less dense than a solid. Books are solids and contain a lot less air. This makes bubble wrap less dense than books.

Water pushes up on boat.

Gravity pulls down on boat.

Buoyant Force

A force is a push or pull on matter. The upward push that keeps an object floating is called buoyant force. To understand the buoyant force that makes things float, you first have to understand gravity. You already know a lot about **gravity**. You see and feel the effects of gravity everywhere every day. Gravity is the force that holds you, and all objects, on Earth. It is a force, or pull, between any two objects. All objects have this pull toward other objects. The pull between most objects is small, and unless an object has a lot of mass, you do not feel its pull.

gravity: a pull between two objects. Gravity is the force that holds all objects on Earth.

When one (or both) of the objects is very massive (which means it has a lot of mass and, therefore, has a lot of matter), you can experience gravity's effects. Earth is very massive, and gravity is the force that pulls everything down toward the center of Earth. Because of gravity, almost everything—people, furniture, trains, and dogs—stay put on top of Earth's surface. In your activity with boats, Earth's gravity pulls on the water and keeps the water in the bucket in which you are floating your boats. Gravity also pulls down on the foil boat.

In designing a boat, an important consideration is why some boats stay afloat, while others do not, and sink. This is a question of how much buoyant force the boat produces.

△ Guide

Introduce gravity by explaining it is the pull between objects with mass. Scientists do not completely understand what gravity is, but they know every object with matter attracts every other object with matter. This attraction between two objects is very small, so when two apples are placed near each other on a table, the pull between them is not noticed. When one object has a lot of matter (when it is massive), the pull between the objects is great enough to notice. Earth has a lot of mass, so we notice the pull between Earth and objects near it, such as falling apples.

If the boat sinks through the water, it is Earth's pull on the boat. The water does not push up enough on the boat to keep it afloat.

TEACHER TALK

"What causes a boat or any other object to sink? A boat sinks because there is a pull between the boat and Earth that pulls the boat toward Earth. If the boat is resting on a table, then it won't sink through the table because the table supports it. If the boat is in the water, the water may not be able to support it, and it may sink."

Explain how water's buoyant force is the push upward on the molecules of the boat by the molecules of water. Sometimes the shape of the boat makes a difference because shape affects how many water molecules are pushing on the boat.

TEACHER TALK

"Imagine we make two boats out of identical materials, each having the same mass. The pull down by Earth on each boat is the same. One of the boats is shaped like a rectangle or a raft, with a large surface that goes in the water. The other is shaped like a sphere or a ball. If a boat has a large surface that interacts with the water, then a lot of water molecules push up on it. If the boat has a smaller surface area, then it won't have as many water molecules pushing up on it."

Demonstrate how two objects with the same mass can have different buoyancy by crumpling a 5" X 5" square of foil into a tight ball and setting it next to a flat 5" X 5" square of foil on a desk.

Consider having students do the same with two squares of foil. Point out that each part of the flat sheet of foil is in contact with the table's surface, so more of the table helps to support the foil. Only a small part of the sphere is touching the table, so the entire sphere is only supported by a small part of the table. Explain how this is similar to objects in water. Objects in water may be supported by few molecules of water or by many.

You could let students know that the surface of the water has special properties that lead to surface tension and that this surface tension adds to the overall upward force on the boat. You should mention that this force is also present, but students will not be considering it when designing their boats.

Gravity pulls things toward the center of Earth, but objects do not continue falling toward the center of Earth. The ground, or other surfaces, resists Earth's pull. In the boat-building challenge, the molecules of the water push up on the molecules in the foil boat at the same time that gravity pulls down on the boat. If the buoyant force of the water pushing up on the boat is as strong as the force of gravity pulling the boat down, the boat will float.

The force pulling the boat down is gravity and the water's buoyant force is the upward push helping to keep the boat afloat.

You may have thought that heavy objects sink and light objects float. But some of you might have gotten the heavy keys to float by shaping the boat in different ways. That shows that weight is not the only factor determining if objects float or sink. To illustrate what is happening in the water, look at the way gravity pulls on something that is not in water.

Crumple one of the 5-inch squares of aluminum into a ball, squeezing out as much air as you can. If you place it on a tabletop, you can see that all of the mass of the foil is pushing down on a very small part of the table. Set another 5-inch square of foil flat on the table, and the same mass of foil is now pushing down on a much larger area. The flat piece of foil touches more of the surface of the table. The piece crumpled into a ball touches less of the table's surface. The mass of the foil ball is concentrated into a smaller area of the table, and fewer molecules that make up the table can push back on it.

These children are able to float in the water because the gravity (downward push) of their bodies is equal to the buoyant force (upward push) of the water. Since a flotation device is less dense than the child, it causes a decrease in the overall density of the person wearing it. This means that less of an upward push by the water is needed to keep the wearer afloat.

NOTES

If you were to place the foil sheet and the foil ball in a bucket of water, what do you think would happen to each? The flat sheet would float. The ball would sink (if *all* the air in the ball had been squeezed out of it). The foil ball would sink because the small area of water in contact with the full mass of the foil does not put enough buoyant force on the foil to keep it above the water. Instead, the water molecules simply slide around and over the foil ball, and it sinks.

When the foil is spread out flat, more of its surface has contact with the water. The same amount of mass from the foil pushes down on a much larger area of water. This creates a situation in which more molecules that make up the water can push up on the foil. As long as the force of gravity pushing down from the foil is equal to the buoyant force pushing up from the water, the foil will float. The flat piece of foil is better able to float because more molecules of water can apply their upward buoyant force to push up on the foil.

The air that fills the open parts in the boat decreases the boat's overall density, making it possible for the buoyant force (upward push) of the water to keep it afloat.

Density and Buoyancy Force

Buoyant force and density work together to affect whether or not something will float. When a boat sits in water, it pushes some of the water away, or displaces it. The water that was pushed away has a certain density. If the boat, including the air in it, is less dense than the water it pushes away, the boat will float. If it is denser than the water it pushes away, the boat will sink. As the density of any object increases, it sinks lower into the water, always displacing an amount of water equal to its weight. The weight of the water that is pushed away, or **displaced**, by the boat is equal to the weight of the boat.

displace: to take the place of.

This is a complicated idea to think about. You will have more opportunities to investigate the effects of gravity, buoyant force, and density. For now, think about your challenge. You are trying to figure out how to make the weight of the keys spread out over a large enough area of the surface of the foil boat so that the buoyant force of the water can keep it afloat. Even as you add more keys, the boat will stay afloat as long as you can find ways to spread the weight of the boat and keys over a larger space.

Density and Buoyancy Force

10 min.

Students discuss how density and surface area affect buoyancy.

META NOTES

The buoyant force exerted on an object by a fluid is equal to the weight of the fluid that the object has displaced. This is the same as Archimedes' Principle: the force pushing up on an object partially or fully immersed in a fluid is equal in magnitude to the weight of fluid displaced. Students do not need to remember this.

△ Guide

Lead students to think about how density and surface area affect the buoyant force on a boat. A boat sitting in water displaces, or pushes away, some water. In order to float, the boat must be less dense than the water it displaces. For an object to float, the buoyant force must be equal to the pull down on the object by Earth's gravity.

TEACHER TALK

"For an object to float, the buoyant force on the object must be equal to the object's weight.The surface area of the object and the density of the object affect whether the buoyant force is equal to the weight. By spreading out the object (the boat and its load), more water molecules push up on it, increasing the buoyant force. By making the object (boat and keys) less dense, you increase the buoyant force on it. Increasing surface area and decreasing density make the boat less dense than the displaced water.

Think about an inflatable toy in a pool. If you filled the toy with air, it would rise to the top of the pool, because Earth's gravity pulls the water more than the air-filled toy and the water pushes the toy out of the way as it goes to the bottom of the pool. If you filled it with sand it would sink because Earth pulls down on the sand more than the water pushes up on it (sand is denser than water).

Could you make the sand-filled toy float? What if you changed the surface area of the toy?"

META NOTES

Where you place the keys in the boat can affect the stability and may cause your boat to tip over and sink. If the keys are all in one corner of the boat, then that corner will feel a greater pull down on it from Earth's gravity. This will cause it to tip over and sink. If there is a hole in your boat, water will fill in your boat, increasing the overall density of your boat and causing it to sink.

If you decide to demonstrate the foil ball and foil square in water, now is the time to do it. Be aware that air trapped in the ball of foil may complicate the demonstration and make it confusing to students. Both the sheet and the ball will probably float in water, because the ball contains air. It is difficult to crumple foil and get all the air out. If you push both to the bottom of the bucket, the rectangular piece will stay at the bottom and the sphere will probably rise to the top again. The foil is denser than water, but the foil ball with air in it is less dense than water.

Emphasize that for a boat to float, the overall push up from the water on the boat must be equal to the pull down on the boat by the water. These pushes and pulls can be adjusted by changing the density of the boat and the surface area of the boat.

NOTES

Reflect

You are going to get another chance to design a boat. You will use the same materials. Think about how your group could design your next boat to better meet the challenge by considering what you now know about gravity, density, and buoyant force. Answering the following questions should help.

1. Think about some of the boat designs that held the most keys. What decisions did the students who designed these boats make that improved the buoyant force of the boat?
2. Did your boat float? If it did not float, why do you think it sank? Discuss *buoyant force* and *density* in your answer.
3. How could you make the boat better able to stay afloat? Remember that you have to float six keys. Use what you have learned about gravity, buoyant force, and density to answer this. Also, take advantage of what you can learn from other groups' designs.

What's the Point?

All objects are made of matter, which is made up of atoms. Atoms combine with other atoms, becoming molecules. All matter, including water and air, is made up of atoms and molecules. Density is the amount of matter in a given amount of space. It is a measure of how tightly the matter is packed together in the space.

For something to float, the force (gravity) pulling down on it cannot be greater than the force (buoyant force) pushing up from the water. To increase the buoyant force pushing up on a boat, you can spread the mass of the object over a greater area of water. This is similar to placing both the flat piece of foil and the crumpled ball of foil in a bucket of water. The flat piece floats, while the crumpled piece, if all the air has been removed, will sink.

When the foil is spread out, more of its surface has contact with the water. The same amount of mass from the foil pushes down on a much larger area of water. More molecules that make up the water can push up on the foil, or any size boat. The greater the surface of an object that touches the water, the more molecules of water can apply their buoyant force and push up the object.

Reflect

10 min.

Students reflect on how the science of boats can help them build a better design.

META NOTES

It is important for students to record their group answers and keep track of their group's ideas and reasons for those ideas and refer back to during the class discussion and during the next section.

⬡ Get Going

Point out that the reading and discussion included many science concepts that can be used to help them build a better boat. Let groups know they will work the way scientists work, discussing their ideas for each question and developing their best group responses. Emphasize that each student is responsible for contributing to the group discussion and recording the group response on his or her own paper.

Tell groups how much time they have to answer the three questions. They should have no more than 10 minutes.

△ Guide and Assess

Lead a class discussion of the *Reflect* questions. During the discussion, listen for the following responses and guide students to these responses if they are having difficulty.

1. Students should mention the designs tested previously that met the criteria and constraints of the challenge. They should describe the surface area and possibly the density. A boat with higher sides keeps more air in and water out, but it decreases the surface area of the boat. A boat constructed with foil that is crumpled rather than folded might hold more air within the body of the boat, making it less dense.
2. Students should describe how the density and surface area of their boat affected the buoyant push up, and how this affected whether their boat sank or floated.
3. Students should describe ways to decrease the density of their boats and to increase the surface area contact with the water.

Wrap up the discussion by letting students know that they will be redesigning their boats in the next section.

NOTES

Assessment Options

Targeted Concepts, Skills, and Nature of Science	How do I know if students got it?
Science and engineering are dynamic processes, changing as new information becomes available.	**ASK:** How did knowing more about the science of boats affect your design ideas? **LISTEN:** Students should indicate they have ideas about improving their boat design based on increasing the surface area of their boats and/or decreasing the density of their boats.

Teacher Reflection Questions

- What difficulties did students have with the concepts of mass, buoyancy, and density? What ideas do you have to assist students next time to apply the science knowledge in this section?
- One of the goals in this section is for students to realize that they are better able to reach their goals if they have more information. Scientists research their questions before they start designing an experiment. What difficulties did your students have in connecting the importance of science knowledge about buoyancy and density to building a better boat?
- How did you gauge the participation in the class discussion? How well did students participate in the wrap-up class discussion? Did each group member add to the discussion? What ideas do you have for next time?

NOTES

NOTES

SECTION 1.4 INTRODUCTION

1.4 Design

Build a Better Boat II

◀ ***1 class period***

A class period is considered to be one 40 to 50 minute class.

Overview

Students apply what they learned about the science of buoyancy as they design and build another boat. This time, there is an additional constraint of making the boat carry more mass. While designing, building, and presenting their new boats, students practice the social practices of scientists by working in small groups and then sharing their ideas with the class to build on each other's ideas.

Targeted Concepts, Skills, and Nature of Science	Performance Expectations
Science and engineering are dynamic processes, changing as new information becomes available.	Students should apply what they now know to a new design and describe how their design and ideas evolved through each iteration.
Scientists often work together and then share their findings. Sharing findings makes new information available and helps scientists refine their ideas and build on others' ideas. When another person's or group's idea is used, credit needs to be given.	Students should work effectively together to redesign and build their boat. Students should listen to each other, try out a variety of ideas, and prepare reasonable presentations in which they share their ideas.
Criteria and constraints are important in design.	Students should identify, describe, and apply all criteria and constraints.
Scientists must keep clear, accurate, and descriptive records of what they do so they can share their work with others and consider what they did, why they did it, and what they want to do next.	Students should refer to their records and describe in detail what they did and why.

Materials	
1 per class	Class list of criteria and constraints
1-4 per group	5" x 5" squares of foil
8 per group	Keys (can be substituted with 1— $\frac{1}{2}$ " hex nut and 2 keys)
1 per group	Stopwatch Bucket or container of water Paper towels, to clean up spills Scissors Masking tape Stream table
1 per student	*Boat Records* page *Solution-Briefing Notes* page

Watch out for spills – they should be cleaned up right away to avoid slipping.

Activity Setup and Preparation

Cut out 5" × 5" foil squares for each group. You will need about four per group.

Homework Options

Reflection

- **Science Content:** How did you use ideas about density and surface area in your new design? Why? *(Students are expected to use the ideas of density and surface area and how they affect how things float in their new design. Students should discuss how they used these ideas.)*
- **Nature of Science:** How did working in small groups and then sharing ideas with the class help to increase your knowledge design a good boat? *(Students should recognize that knowledge is a constraint or limitation. That is why scientists share their ideas in groups and then with the larger scientific community, so they can build upon each other's ideas.)*

Preparation for Back to *the* Big Question

- **Nature of Science:** How do you think scientists work together to solve problems? How do you think this is similar to how you worked on the boat challenge? *(Look for descriptions of scientists working in groups and presenting their ideas to the scientific community. Students should support their ideas with examples of how they worked on the boat challenge.)*
- **Nature of Science:** Write a story about a team of scientists working on a project, describing how they work together. *(Look for descriptions of scientists working in small groups and then in larger groups to build upon each other's ideas.)*
- **Nature of Science:** Draw a diagram representing your interactions with other members of your class while you were working on the boat challenge. Describe what you were doing during each interaction. Draw a diagram of how you think a scientist interacts with other scientists when working on a project. *(Students should draw a diagram showing interactions individually or in small groups connecting to larger groups, going back to the individual or smaller group and connecting again with larger groups.)*

NOTES

*1 class period** ▶

1.4 Design

Build a Better Boat II

5 min.

Introduce students to the new challenge—to design and build an improved boat with the modified criterion of making it carry eight keys.

SECTION 1.4 IMPLEMENTATION

1.4 Design

Build a Better Boat II

You now have the knowledge and experience to design a boat that will hold six keys. However, just when you thought you had it all figured out, the gatekeeper tells you that your friends will need two more keys to turn off the alarms. He wants you to show him a final design that will now carry eight keys and an explanation of why you think this will work. He will then be confident in giving you the keys to the fort.

You will discuss as a class what you think will be the best design to support the additional weight of two more keys long enough to float them into the fort. As you discuss the boat design, keep in mind what you just read about density and buoyancy. Use this knowledge to help you design the right boat.

DIG 23 DIGGING IN

◯ Engage

Remind students of the original challenge and introduce the new criterion. Emphasize that making the boat carry eight keys is necessary to turn off the alarm and is essential to meeting the challenge.

*A class period is considered to be one 40 to 50 minute class.

TEACHER TALK

"You know a lot more about how to make a boat that can hold a lot of mass and still float. That is a good thing, because now the gatekeeper has told you that you can't get your friends out unless your boat can carry eight keys. So your task is to make a new boat using a 5" × 5" piece of foil that can carry eight keys for 20 seconds. You have 10 minutes to design it."

NOTES

Update Your Criteria and Constraints

Now that the challenge has changed, review the list of criteria and constraints. Update these lists. Then consider these changes as you design and build your new boat.

Plan, Build, and Test Your Design

Materials

- 5" × 5" square of foil
- 8 keys
- bucket of water

As you design your new boat, you are welcome to use ideas other groups have developed, but you must make sure to give them credit. Once again, be sure to keep track of the design and the number of keys the boat holds. Record your results on a *Boat Records* page.

After completing your boat, your group will present your new solution to the rest of the class. During your presentation, have your *Boat Records* pages handy to report your results to the class. Good luck!

Communicate Your Solution

Solution Briefing

Now it is time to share your new boat with the class. Once again, you will participate in a *Solution Briefing*.

As before, spend some time preparing for your presentation. Be prepared to answer questions such as the following:

- How is your design constructed?
- What materials did you use, and how much of each?
- Why did you build it the way you did?
- How does the design meet the criteria?
- How did the constraints affect the design?
- What past experiences helped you make your design?
- What science knowledge helped you make your design?
- What problems remain?
- What else do you want to test?

Update Your Criteria and Constraints

5 min.

The students update the class list of criteria and constraints based on what they have learned about density and buoyancy and the new criterion of carrying eight keys.

△ Guide

Now that a new criterion has been given, the class needs to update their class list of criteria and constraints. Ask students what needs to be added or changed and update it accordingly. Students should add that the boat needs to carry eight keys to the list of criteria. They may also want to add something about density and buoyancy or edit the criteria they have already listed.

Plan, Build, and Test Your Design

10 min.

Students design and build a boat, applying the new criteria and constraints and what they now know about the science of boats.

◇ Evaluate

Make sure that the criteria include that the boat needs to carry eight keys.

⬡ Get Going

Begin by reminding students of the importance of planning the design before building it, and the usefulness of building the design to see if it will work or what needs to be improved.

Distribute materials including the *Boat Records* page and let students know they have 10 minutes.

△ Guide

While groups are working, check to see how they are doing. Most groups should not have difficulty with creating a design since this is the third time they have designed and built a boat. They may still have trouble writing the reasoning behind their design choice. Ask guiding questions such as, "Why did you choose to fold the corners up like that?"

□ Assess

Monitor groups' progress. If it looks like the majority of the class needs more time, then give groups extra time.

Communicate Your Solution

Solution Briefing

20 min.

Groups present their designs, discuss their results, and keep track of other groups' designs and results.

△ Guide

Transition students from building their design to preparing to share it by reminding them that during a *Solution Briefing,* ideas are shared and advice is given on how to improve the design.

TEACHER TALK

"You've designed another boat. Now it is time to share ideas during a *Solution Briefing*. Remember that during a *Solution Briefing,* we share our ideas to get advice and new ideas on how to improve our design."

Remind students that the questions will help them prepare for their presentation. They will also listen for design ideas from everyone's presentations and will keep track of how designs work using the *Solution-Briefing Notes* pages.

TEACHER TALK

"It is important that you all think about what you are going to say when you present your designs. There is a list of questions that will help you think about what to say during your presentation in your texts."

⬡ Get Going

Give groups five minutes to prepare their presentations.

After groups have finished preparing, remind students they will need to ask questions. If they do not understand a design and how it works, they should ask the presenting group to clarify. Remind students that they should also voice their opinions and ideas, but in a polite and considerate manner, using language such as "I didn't hear the answer to ...", "Could you clarify .... for me?", "I agree with ... because..." or "I disagree with... because." Reasons should be given.

Begin the presentations. Give each group two minutes to share and then two minutes for questions.

Remind the class that they should fill out their *Solution-Briefing Notes* page.

△ Guide

During the discussion, highlight how students used the ideas about density and buoyancy to improve their designs and how iteration and collaboration have been important in the design process.

While groups are presenting, you and the class should be listening for answers to the questions listed in the student text. If any of these items are not answered during the discussion, give the class time to ask questions to obtain the answer. Otherwise, guide the presenting group to providing an answer using the ideas below. Listen for:

- Students should describe their design, pointing out its construction.
- For this challenge, everyone will have used foil as this is one of the constraints.
- Students should state the reasons for their overall design and for some of their specific design features. This may include how the criteria and constraints were met, but that question is also asked later so students may state it later.
- Students should describe how their boat meets the criteria. They may have already answered this when discussing how well it works.
- Students should describe how each constraint affected their design and how the requirement to carry eight keys affected their design.

- Students should discuss their past boats and how learning science knowledge affected the design. Point out if any of the students use others' examples without giving credit. Remind students to give credit.
- Students should discuss what they learned about how some objects float in water and how an object's density affects how much it floats in water.
- Students should describe any problems they feel still remain and ideas they think might be good solutions to these problems.
- Students should state what other ideas they have.

NOTES

Reflect

10 min.

Lead a class discussion on how students improved their designs by applying ideas of density and buoyancy with a new criterion of carrying more mass. Build up to the importance of iterative design, collaboration, and building on the ideas of others.

When you present your boat design, your group will need to justify the design decisions you made and share the results of any other designs you created. As before, keep notes on a *Solution-Briefing Notes* page. As you listen to the presentations, remember to ask questions if anything is unclear.

Reflect

Answer the following questions. Be prepared to discuss your answers with your class.

1. Review the criteria and constraints for the first *Boat Challenge* and then for the second *Boat Challenge*. Which criteria and constraints are different in the second challenge?
2. What changes did you make in your design to address the new criteria and constraints?
3. How did you change your original design to include the science of density and buoyancy?
4. What criteria and/or constraints were you unable to meet? Why?

What's the Point?

Now that you have designed your boat two times, you have seen how useful iteration is. Each time you iterated on your design, you had a chance to use what you learned from the last time. Each time, as a result of using new knowledge, you made your boat better.

Sometimes the new ideas you had were based on new science you learned, and sometimes you learned from what other groups had done. You may have remembered experiences that helped you form ideas. Ideas can come from all of these places. It is important, when ideas are borrowed from others, to give them credit. This is how the fields of science and engineering make progress. Also, people feel good when others use their ideas and give them credit.

⬡ Get Going

Let groups know how much time they have (about five minutes) to answer the questions and prepare for a class discussion.

☐ Assess

While groups are answering the questions, monitor their work and check their understanding of the questions. Decide how you will focus the discussion based on students' responses. If students seem to understand these questions, lead a brief discussion of them.

△ Guide

Begin with a discussion of how designs changed in response to science knowledge (density and buoyancy) and new criteria and constraints (affordability and cost of materials). Transition the discussion to explore how the boat challenge is representative of challenges in general and what students have learned about iterations, collaboration, and building on the ideas of others.

NOTE: Students may have already brought up some of the answers to the questions in the *Solution Briefing*.

Listen for the following responses to the questions in the student text:

1. Students should list the criteria and constraints they added during this section, and describe any other changes they made to the list of criteria and constraints.
2. Students should describe how they changed their designs to meet the new criteria and constraints. They should explain how these changes meet the criteria and constraints.
3. Students should identify the ways they changed their designs to make use of new science knowledge. They should describe how the changes make use of the new science knowledge.
4. Students should list any criteria and constraints they could not meet and the reasons why they could not meet them. Ask these groups what ideas they have that could help them meet the criteria and constraints.

Teacher Reflection Questions

- How effectively are students analyzing design challenges in terms of criteria and constraints? What can you do to help them with this?
- Just as it takes iterations to improve a boat, it also takes iterations to improve the social practices used throughout *PBIS*. How did students' abilities to participate in the *Solution Briefings* improve in this section compared to the *Section 1.2?* What do they still need to work on?
- What time-management issues occurred during this *Learning Set?* What ideas do you have for next time you teach this section?

BACK TO THE BIG QUESTION INTRODUCTION

Back to the Big Question

How do scientists work together to solve problems?

Overview

Students identify the practices of scientists that they used in this *Learning Set* and consider how these practices enabled them to meet the challenge of the *Learning Set.* They see that the practices of scientists involve collaboration, helping them to begin to answer the *Big Question: How do scientists work together to solve problems?* Students define or describe each of the practices of scientists they identified, connecting what they have learned from the challenge to these concepts.

Targeted Concepts, Skills, and Nature of Science	Performance Expectations
Scientists collaborate in their work and then share their findings. Sharing findings makes new information available and helps scientists refine their ideas and build on others' ideas.	Students should identify collaboration as an important part of the practices of scientists.
Criteria and constraints are important in design.	Students should identify the practice of listing criteria and constraints as an important practice of scientists.
Scientists must keep clear, accurate, and descriptive records of what they do so they can share their work with others and consider what they did, why they did it, and what they want to do next.	Students should identify record-keeping as an important practice of scientists.

Targeted Concepts, Skills, and Nature of Science	Performance Expectations
Scientists often work together and then share their findings. Sharing findings makes new information available and helps scientists refine their ideas and build on others' ideas. When another person's or group's idea is used, credit needs to be given.	Students should identify ways they built on other groups' ideas and identify this as an important practice of scientists.

Homework Options

Reflection

- **Science Process:** Which of the changes that you made to your boat design do you think made the biggest difference? How did they make a difference? *(Look for ideas that students can generalize and use to address the challenges in the next* Learning Sets*.)*

NOTES

Learning Set 1

Back to the Big Question

5 min.

BACK TO THE BIG QUESTION IMPLEMENTATION

Learning Set 1

Back to the Big Question

How do scientists work together to solve problems?

Over the past few days, you and your classmates have been working to create a boat that can carry six or eight keys for at least 20 seconds. The last boat you built was probably a lot better than the first one. During this activity, you took part in several practices that scientists use when they solve problems. Think about some of the things you did in this *Learning Set*.

You identified the criteria and constraints of your challenge. Criteria are the requirements your solution must meet. Constraints are the factors that put limits on your solution. You also saw how criteria and constraints could change as you attempt to solve the problem.

You learned that there is a difference between copying and building on the ideas of others. You saw some designs of other groups that may have looked very good. In your next attempt, you may have used some of these ideas. Others might have used some of your ideas. This is how scientists work and how science grows as a field. Science builds on the ideas of others.

Scientists work together. They support each other. Working together to build ideas and understanding is called collaboration. In this class, you will collaborate to solve problems or meet challenges. As you collaborate, you will share ideas with others. Others will share ideas with you. One way you collaborated was to participate in a *Solution Briefing*. Scientists often present solutions or ideas while they are trying to solve problems.

Iteration can help you achieve a challenge or solve a problem. You probably saw that it is not always easy to achieve success the first time you try something. But once you shared and saw the ideas of others and learned some scientific concepts, you were able to plan and build a better boat. Scientists also use iteration when solving problems.

Project-Based Inquiry Science DIG 26

△ Guide

Use the student text and the class's experience with the boat challenge to help students begin to link their ideas about criteria and constraints, collaboration, building on others' ideas, and iteration to answer the *Big Question: How do scientists work together to solve problems?*

Discuss the importance of identifying criteria and constraints and why scientists need to identify the criteria and constraints when solving problems.

Discuss collaboration and copying versus building on each other's idea. Scientists collaborate—they work together and share their ideas in the same way the students did and then they build on each other's ideas. Scientists give credit when they build on someone's ideas.

Discuss iteration—when scientists begin working on a problem, they know they may not be successful at solving it the first time and may improve it the more they work on it.

NOTES

Reflect

5 min.

Back to the Big Question

In this *Learning Set*, you probably approached problem solving differently than you have in the past. You now have a better idea of how scientists work together to solve problems.

Reflect

Define or describe the following ideas and why they are important to working together as student scientists. Be prepared to discuss your answers with your class.

- criteria and constraints
- iteration
- collaborating
- building on the ideas of others
- record keeping
- asking questions
- using science knowledge

DIG 27 DIGGING IN

◇ Evaluate

You can have students write their definitions, descriptions, and why these things are important or you can bring them up as part of the class discussion. Look for the following:

- The criteria are the goals that the design should achieve; the constraints are the limitations on the design. Listing the criteria and constraints at the beginning of the design process helps understand the challenge.

- An iteration is a repetition of the design process. Designers often have to go through many iterations to meet the criteria and constraints of a challenge.
- Collaborating is working together. In the design process, it is helpful to share ideas and to get feedback from your peers.
- When you build on the ideas of others, you use what they have discovered in your own work and credit them. Building on the ideas of others allows scientists to make great progress and ensures that everyone receives credit for their contributions.
- Keeping records means to write or draw what you are doing. It allows you to share your ideas with others and remember what you did and why you did it.
- When listening to others' ideas or findings, it is important to ask questions to clarify and to make sure you understand. This can also help others make sure that their ideas make sense and their explanations are complete.
- Science knowledge helps understand why some features work better than others, and to use these features as effectively as possible.

META NOTES

Remember students will not have a complete answer to the question: ***How do scientists work together to solve problems?*** until the end of the Unit. Students will continue throughout the Unit to be introduced to and practice the nature of science and its social practices and processes.

NOTES

NOTES

LEARNING SET INTRODUCTION

Learning Set 2

The Lava Flow Challenge

◀ *4 class periods*

Student pairs design and then redesign a procedure for determining the rate of lava flow using dish soap and a plastic plate. The importance of uniform procedures is highlighted when students' initial data are inconclusive.

A class period is considered to be one 40 to 50 minute class.

Overview

Students are introduced to procedural design and methods of data analysis. They are challenged to develop a procedure for measuring lava flow accurately using dish soap and a plastic plate to model lava on a slope. The class devises an initial procedure for determining how fast dish soap flows across a plate. Students run the procedure in pairs. Once students have gathered data from their simulations, the class compiles their results in bar graphs. By organizing data in bar graphs, the differences in results become obvious, leading students to examine their procedures. As a class, students identify the differences in their procedures. They discuss the importance of specificity in procedures for obtaining repeatable results. Then they design a more specific procedure for addressing the challenge. After plotting their data again, they see that their results are more consistent and informative when they use clear, specific procedures. By revising and repeating their investigation, students see the benefits of a well-designed procedure.

LOOKING AHEAD

This *Learning Set* was designed so that students complete *Sections 2.1* and *2.2* in one class period. Try not to break up the data collection in *Section 2.2*, as it will alter students' analyses. It would be ideal to complete the whole ivestigation in one class period.

Targeted Concepts, Skills, and Nature of Science	Section
Scientists often collaborate and then share their findings. Sharing findings makes new information available and helps scientists refine their ideas and build on others' ideas. When another person's or group's idea is used, credit needs to be given.	2.3, 2.4
Scientists must keep clear, accurate, and descriptive records of what they do so they can share their work with others and consider what they did, why they did it, and what they want to do next.	2.2
Graphs are an effective way to communicate the results of scientific investigations.	2.2

Targeted Concepts, Skills, and Nature of Science	Section
Identifying factors that lead to variation is an important part of scientific investigations.	2.3
Scientific investigations and measurements are considered reliable if the results are repeatable by other scientists using the same procedures.	2.2, 2.3, 2.4
When volcanoes erupt, magma reaches Earth's surface and is called lava. There are several different types of lava.	2.4
Scientists use models to simulate processes that happen too fast, too slow, on a scale that cannot be observed directly (either too small or too large), or that are too dangerous.	2.2, 2.4

Students' Initial Conceptions and Capabilities

- Most students will have a good idea of what a scientific investigation is and how it works, but they may not realize that investigations are usually carried out within a scientific group, and the results shared among the scientific community. (Fort & Varney, 1989; Newton & Newton, 1992; Mead & Metraux, 1957.)
- Students may also initially be willing to accept conclusions based on very weak evidence. (Wollman, 1977a, 1977b; Wollman & Lawson, 1977.)
- Identifying all variables involved in an investigation may be especially difficult. (Linn & Swiney, 1981; Linn, et al. 1983)
- Students may think of models as physical copies or reality rather than conceptual representations. (Grosslight, Unger, Jay, & Smith, 1991.)

Understanding for Teachers

Carefully designed and performed scientific procedures are critical to the reliability of scientific evidence. Scientists perform, validate, and revise their procedures to account for procedural and data collection errors. The iterative process allows scientists to improve procedures and collect reliable data.

Scientific claims must be supported by evidence. This puts stringent criteria on what makes an experiment valid. For a scientific test to be considered valid, it must be repeatable. When many scientists perform the same test using the same procedures, they should get the same results, within an acceptaple range of error.

No measurement is exact. Every measurement is limited to the scale of the measuring device and the ability of the observer to read the scale. Imagine taking a measurement of length using a meter stick. It would be difficult to accurately measure to the nearest ½ millimeter because a millimeter is 1/1000th of meter. Many students believe machines can make exact measurements. A digital scale may give the mass up to the nearest 1/100th of a gram each time you measure the same mass, however, you cannot be certain of any value smaller than 1/100th of a gram. Many machines introduce errors because the machines have moving parts. These parts have friction and associated temperature changes that make measurements more difficult. The range of data determines the precision of the data. If the range is small, the results are more precise. Precise measurements are considered reliable because they are consistent. Many trials are completed to determine the precision of the measurement. However, the measurements may not be accurate. Accuracy is how close to the actual value the measurements are. A machine measuring temperature may consistently give you values that are 10°C too low because it is not calibrated correctly. In this case, the measurements are very precise, but not very accurate.

In science, if there is a single valid test (one that has been repeated and accepted by many scientists) that goes against a claim, then the claim itself is no longer valid and needs to be revised or removed. Newton's law work very well for speeds much less than the speed of light, but break down when objects begin moving very fast. Einstein's theories are considered more adequate at describing interactions with forces.

In this *Learning Set*, students experience how scientists design procedures to provide reliable results and how to determine reliability. Students identify that their results are more reliable if results from doing the procedure many times are similar. Students use bar graphs to communicate and interpret results, but they do not need to master bar graphs by the end of this *Learning Set.*

META NOTES

The goal of this Learning Set is to support students in learning about the impact of procedures on the quality of data collected. The motivational context of the *Learning Set* asks students to make a recommendation to a Hawaiian town. It is not expected that students should learn about all types of lava and make a recommendation that uses the soap as a model for the lava flows. Rather, the context is motivational. The intention of this *Learning Set* is for students to analyze the connection between carefully designed and performed procedures and measurements and the reliability of collected data.

Constructing a Bar Graph with Students

Students are able to see the range of the data when they create a bar graph. The bar graph is a class artifact that provides a visual representation of the data. Students will provide all their data points for the creation of the class bar graph. It is important that students are able to create the bar graph on their own graph paper. Provide time for students to complete this part of the activity individually and with the class.

To make a bar graph showing the students' measurements of the flow of dish soap across a plastic plate, begin with a blank graph. Label the *x*-axis (horizontal axis) with the time, in seconds. These numbers will be between 0 and 300 seconds. Observe students as they work to determine the high and low numbers before labeling the columns. When labeling the *x*-axis, begin at the left with a number two or three seconds lower than the lowest number and end at the right two or three seconds higher than the highest number.

Label the *y*-axis (vertical axis) with numbers that correspond to the number of data points for each lava flow time. If there are 15 groups in your class and each group runs five trials, it is unlikely that more than a third of the 75 trials will have the same result, so the largest number on the vertical axis can probably be around 25. See the example below.

Entering data on the line plot is a whole class activity. Ask each group to report their test results and mark each number with an X on the grid. If a group reports 20, place an X above the number 20 on the *x*-axis. If another 20 is reported, place another X directly above the previous X. As the number 20 is called out, the stack of X's grows. One effective way to report the bar graph is to use a transparency of the graph. As you record the data on the graph transparency, students will record the same data on their own line plots.

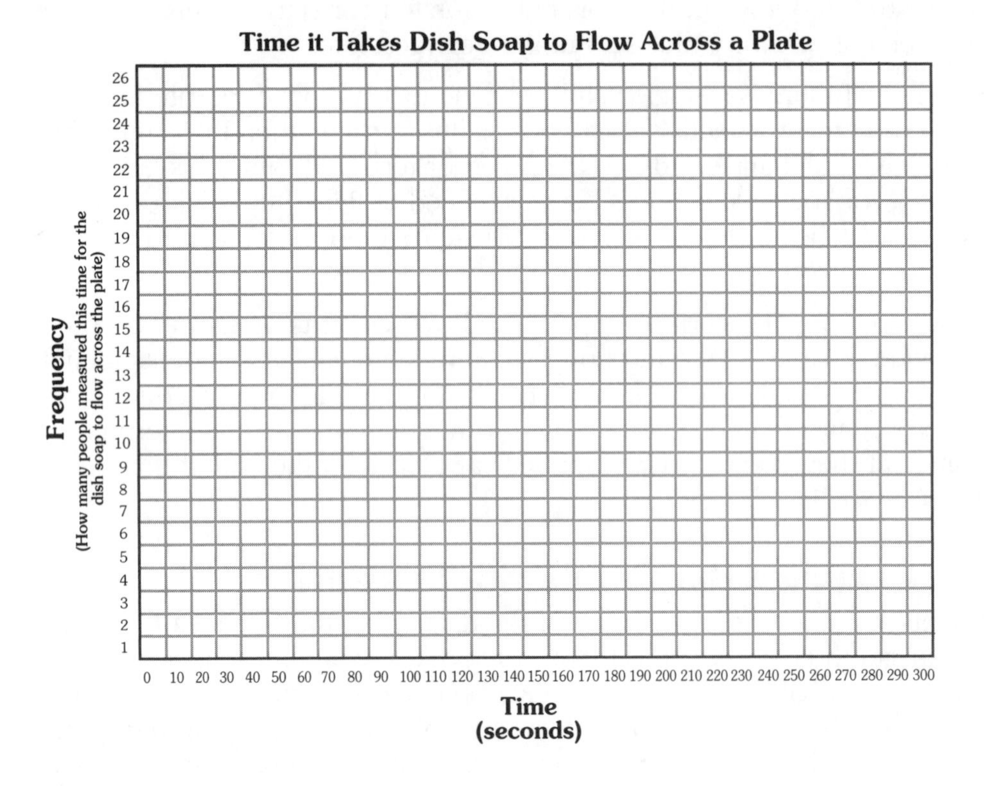

Interpreting Bar Graphs

In this investigation, the range of the data on the bar graphs reveals how well students are controlling their procedure and repeatedly measuring the same event. The initial bar graphs will have a large range (with students reporting trials of anywhere from 0 seconds to 300 seconds), showing few repeating data points (multiple X's in a column). The data will be spread and look similar to the following graph.

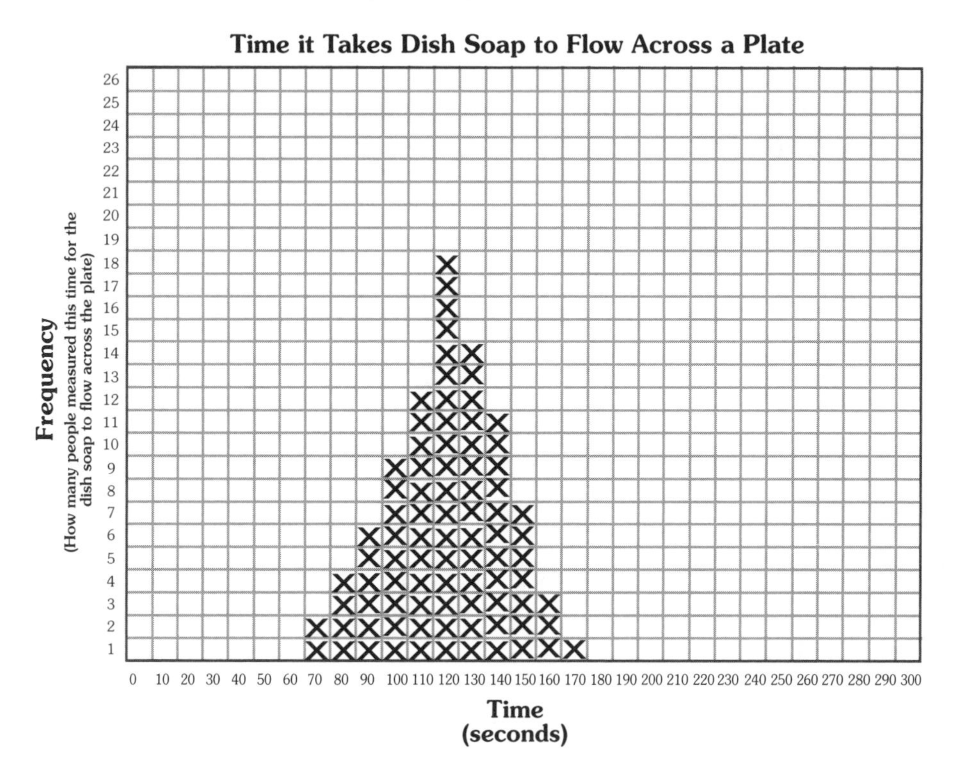

As students revise their procedures, they should see the range of results get smaller. The columns of X's will be more clustered, producing a smaller range of numbers. This bunching of data indicates that students have better standardized their procedure within groups and across groups.

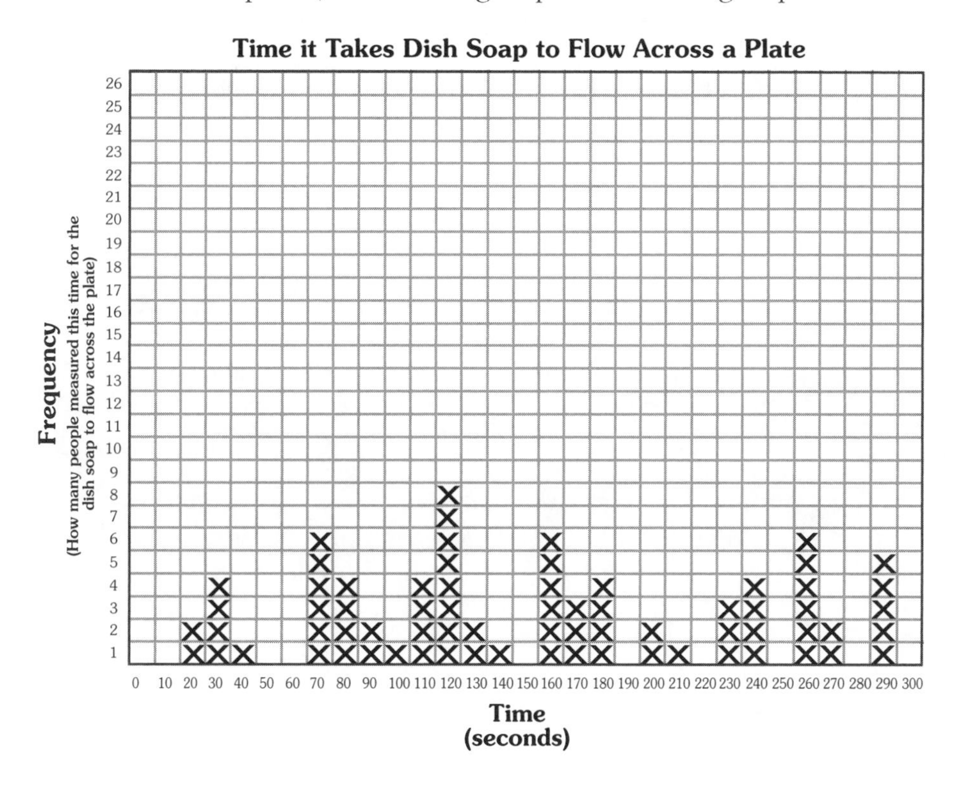

$\frac{1}{2}$ class period* ▶

Learning Set 2

The Lava Fow Challenge

< 10 min.

Students are introduced to the challenge of the Learning Set*: to determine how fast lava flows.*

SECTION 2.0 IMPLEMENTATION

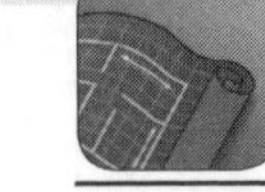

Learning Set 2

The Lava Flow Challenge

volcano: a place on Earth from which melted rock, ash, gases, and other material can escape from beneath Earth's surface.

rock: a naturally formed, non-living solid mass composed of grains of Earth material.

magma: melted rock found beneath Earth's surface.

lava: melted rock that has reached the surface of Earth.

vent: an opening on Earth's surface that allows lava, ash, gases, or other volcanic material to escape.

Nothing shows the power of nature quite like a volcanic eruption. Streams of red-hot lava flowing over land make spectacular photographs. However, these are life-threatening situations. In this *Learning Set*, you will be working as part of a team of scientists. You have been hired by a company that makes measurement equipment. The company wants you to develop an accurate procedure for measuring lava flow. You will have to send the company a procedure along with evidence that your procedure is accurate.

The Hawaiian Islands are volcanic islands. They are made of volcanic material laid down from repeated eruptions. Some of these volcanoes are still active today. Towns near volcanoes usually develop emergency evacuation plans that allow them to move people out of the area safely and quickly if a volcano begins to erupt. Towns usually develop one plan for quick evacuation and other plans for slower evacuations. When a volcano begins to erupt, towns choose the right plan based on how fast the lava is flowing. To determine this, the people of the town need an accurate procedure for measuring the lava flow.

A **volcano** is a place on Earth from which melted **rock**, ash, gases, and other materials can escape from beneath Earth's surface. Volcano is also the name for the mountain created by the hardened rock. Rock, melted deep in Earth, is called **magma**. When the magma flows out of the volcano, it is known as **lava**. Lava and gases escape through an opening on Earth's surface called a **vent**.

When lava is very thick, it moves slowly. However, thick lava is likely to cause explosions because hot gases get trapped in it. When lava is thinner, it moves much more quickly. Thinner lava is more likely to trap people who live in the surrounding area. It is important to figure out a way to accurately measure the rate of lava flow so that townspeople will know how much time they have to evacuate during an eruption.

DIG 28

Project-Based Inquiry Science

○ Engage

The challenge scenario is described in *The Lava Flow Challenge.* The scenario provides students with a context. Make sure they understand their goal is to develop a procedure and measure the time it takes for the dish soap to move across the plate.

*A class period is considered to be one 40 to 50 minute class.

TEACHER TALK

"In this challenge, we need to imitate the flow of lava using everyday materials. You need to develop a procedure for measuring how fast the dish soap will flow. Imagine that the dish soap is lava and think about what you will tell the town."

Provide students with some background information about lava and volcanoes by reading the information in the student text. Students should understand that magma is very thick, hot rock below Earth's surface. Magma is called lava when it flows from a volcano. Lava is hot and varies in thickness. The thickness and the quantity of lava can determine how fast the lava flows.

TEACHER TALK

"Who knows something about how volcanoes erupt? *(Listen to one or two stories if students have them. If not, you may want to tell them a story—you might talk about Pompeii or Mount St. Helen.)*

What is lava? *(Lava is magma that has reached Earth's surface. Magma is very hot, liquid rock that lies below Earth's crust.)*

What is it like? Where have you seen it? What are some things we can say about lava now, before we start investigating?

If no one has discussed types of lava, tell students:

There are different types of lava. Some lava is thick. It moves slowly and is likely to trap gas. The gas gets hot and then explodes, which can cause fires in structures nearby. When lava is thin, it moves quickly and is likely to trap creatures (including humans) that live near the volcano."

You will work with a partner on this challenge. You will be given materials to imitate the flow of lava over the land. You will use a plastic plate, model lava (dish soap), and a stopwatch. You and your partner will develop a procedure for measuring how fast the dish soap moves across the plate. Imagine that this is the same way lava would flow over land. You will share your procedure with your class, and together, the class will develop a procedure that everyone agrees is accurate and reliable.

DIG 29 DIGGING IN

Describe the structure of the activity for students. They will work with a partner using safe materials to imitate the flow of lava over land. Remind students that they will use a plastic plate to imitate the land and dish soap to imitate the lava. They will present their procedure for measuring how fast the dish soap moves across the plate to the class.

◇ Evaluate

Make sure students are ready to start the challenge by checking if they understand they will develop a procedure to determine how fast the soap moves across the plate.

SECTION 2.1 INTRODUCTION

2.1 Understand the Challenge

Identify Criteria and Constraints

◀ $\frac{1}{2}$ ***class period***

A class period is considered to be one 40 to 50 minute class.

Overview

Students have been challenged to develop procedures for measuring how fast dish soap moves across a plate. In this section, they identify the criteria and constraints of developing the procedure and collecting data.

Targeted Concepts, Skills, and Nature of Science	Performance Expectations
Scientists often collaborate and then share their findings. Sharing findings makes new information available and helps scientists refine their ideas and build on others' ideas. When another person or group's idea is used, credit needs to be given.	The class should create a list of the criteria and constraints of the investigation.
When volcanoes erupt, magma reaches Earth's surface and is called lava. There are several different types of lava.	Students should be able to describe the difference between magma and lava.
Scientists use models to simulate processes that happen too fast, too slow, on a scale that cannot be observed directly (either too small or too large), or that are too dangerous.	Students should be able to describe their model that used dish soap for lava and a plate for land, and why they used this model rather than real lava and land.

NOTES

LOOKING AHEAD

This *Learning Set* was designed so that students complete *Sections 2.1* and *2.2* in one class period.

META NOTES

In this *Learning Set*, students learn about the need for standard procedures and consistent measurements. Students are not expected to learn about volcanic activity or lava. The volcanic activity and lava flow information are vehicles for learning about procedures and measurements.

Homework Options

Reflection

- **Science Process:** One of the requirements for this investigation is to run at least five trials. The number of trials helps to show that the time you measured for the dish soap to flow is consistent. What are some things that might make the time measurements inconsistent? How can you make sure they are consistent? *(Students' answers might include: holding the plastic plate at different angles could affect the time it takes for the dish soap to reach the other side of the plate; using different amounts of dish soap can affect the time it takes to reach the other side of the plate; inconsistent or improper use of the stopwatch could effect measurements. In each of these cases, deciding exactly how to run the procedure before beginning should lead to consistent results.)*
- **Science Process:** What lessons from *Learning Set 1* can you apply to this challenge? *(Students' responses might include the need for criteria and constraints, keeping good records, iterations, and collaborative work.)*

NOTES

SECTION 2.1 IMPLEMENTATION

2.1 Understand the Challenge

Identify Criteria and Constraints

Before you get started, make sure that you understand what your challenge is. There are two features of the challenge: the criteria and the constraints.

Remember that criteria are things that must be satisfied to achieve a goal or answer a question. Constraints are factors that will limit how you can go about doing that. Think about and record the goals of the challenge. Think about the limits that have been placed upon you for this challenge. For example, you cannot work with actual lava.

What's the Point?

You have been given a new challenge. Remember, to be successful, you need to understand the parts of the challenge. You need to figure out what you need to achieve (criteria). You must also consider the limits you are working under (constraints). By identifying the criteria and constraints, you are more likely to be successful with your challenge.

The largest of the chain of Hawaiian Islands, the big island of Hawai'i consists of five volcanoes. Several of the volcanoes have erupted over the past 200 years. During some of the spectacular eruptions, lava as hot as 1204°C (2200°F) flows out of the volcanoes.

△ Guide

Initiate a discussion of what students need to do to find out how fast soap flows on the plate. Remind them that two important features of a challenge are criteria and constraints.

2.1 Understand the Challenge

Identify Criteria and Constraints

< 10 min.

To begin the Lava Flow Challenge, *lead students to identify the criteria and constraints of this challenge.*

META NOTES

Analyzing the challenge in terms of criteria and constraints tests students' understanding of the challenge and provides direction for their investigation. In this case, once students identify the criterion of constructing an accurate model of flowing lava to determine how fast the model lava flows, they will understand the goal of their investigation.

It might be difficult for students to distinguish between criteria and constraints. The goal for *PBIS* is for students to begin to see ideas affect scientific investigations. It's not important for them to have committed to memory the differences between criteria and constraints. The focus should be on the concepts rather than the definitions.

TEACHER TALK

"Before we begin, we need to make sure we have a clear picture of our goals and the limits we're working within. Remember how in the last *Learning Set* we called the goals and limits our criteria and constraints.

What are some criteria for this challenge? (*accurately measure time of flow*) What are some constraints for this challenge? (*the materials provided*)"

META NOTES

In *PBIS*, students keep records of their work individually and as a class. When discussions happen as a whole class, a record of the discussion should be created. This record could be written on the board, a poster, a transparency, or a word document projected for students to see. It is important to make these records so that the class can refer to them later.

META NOTES

Students may list the three materials being used in their model as three constraints. This is O.K. Since students are required to use these materials, they become a limiting factor to the challenge.

With the class, identify the criteria and constraints of this challenge. If students are not able to identify criteria and constraints, remind them criteria are goals that must be satisfied to achieve the challenge and constraints are factors that limit how you can solve a problem. It might help to have students think about what they need to accomplish (to determine the time it takes the dish soap to flow across the plate), the criteria of the challenge. You can have students think about the supplies they have available (dish soap, plastic plates, and stopwatches), which are the constraints of the challenge.

As students identify criteria and constraints, record them so that students can refer to the list when they design their procedures in the future.

◇ Evaluate

Review the criteria and constraints list. Make sure students' lists include measuring the time it takes the dish soap to flow across the plastic plate as a criterion and using dish soap to model lava, plastic plates to model the land, and stopwatches as constraints.

Teacher Reflection Questions

- Identifying criteria and constraints is difficult for students. How well have students identified criteria and constraints in this section?
- It is common for students to have difficulty distinguishing between these concepts. How have you helped students begin to distinguish between criteria and constraints?
- When creating the list of criteria and constraints, how did you get students to participate? What ideas do you have for the future?

SECTION 2.2 INTRODUCTION

2.2 Investigate

Modeling Lava Flow I

◀ *$\frac{1}{2}$ to 1 class period*

A class period is considered to be one 40 to 50 minute class.

Overview

Students design and run a procedure to determine how fast lava flows over a landscape. The design of the procedure is constrained by the criteria and constraints identified in the previous section. When each pair's results are graphed, students see a large range of results. This wide range indicates that their results are unreliable. Through a discussion, students realize that their procedures were very inconsistent which led to unreliable results. Students see the need for uniform procedures and measurements, and for collaboration and communication to confirm the reliability of results in scientific investigations.

LOOKING AHEAD

Pacing in this lesson is important. It is easier to complete data collection and analysis in one class period than to break it up over two.

Targeted Concepts, Skills, and Nature of Science	Performance Expectations
Graphs are an effective way to communicate results of a scientific investigation.	Students should draw and interpret the class's bar graphs.
Scientists must keep clear, accurate, and descriptive records of what they do so that they can share their work with others and consider what they did, why they did it, and what they want to do next.	Students should record their procedures and results in detail and describe why this is useful.
Scientific investigations and measurements are considered reliable if the results are repeatable by other scientists using the same procedures.	Students should evaluate the spread of results and their procedure based on whether the results are likely to be repeatable.
Scientists use models to simulate processes that happen too fast, too slow, on a scale that cannot be observed directly (either too small or too large), or that are too dangerous.	Students should be able to describe why scientists use models.

Materials	
1 bottle or cup per pair	Dish soap (model lava)
8-10 per pair	Plastic plates
1 roll per class	Paper towels
1 per pair	Stopwatch
1 per special request	Ruler (Only provide if requested. Do not put on display.) Measuring spoon (Only provide if requested. Do not put on display.)
1 per student	*Lava Flow Data* page Graph paper

Activity Setup and Preparation

Arrange the classroom so students will be able to work in pairs effectively.

Prepare a bottle or cup of dish soap for each pair. Have paper towels ready for students to wipe their plates with. Make sure you have rulers and measuring spoons, but do not display these items. Students should not get these items this time—unless they specifically ask for them.

Homework Options

Reflection

- **Science Process:** What do you think the bar graph would have looked like if everyone had used the same procedure? *(Student should correctly link the idea of standard procedures to consistent results, and consistent results to clustered data.)*
- **Science Process:** Think about your class's results. What results do you think other researchers would get if they were given the same criteria and constraints and then designed their own procedure? *(Students may see that if they have not been able to get consistent results, they cannot say what results other researchers would get.)*

META NOTES

Students should see this question only after they have done the activity, because it reveals information that students should discover during the activity.

- **Science Content:** Consider the following two bar graphs. They represent the weight of candy bars produced in a day by a manufacturing plant before and after they adopted new equipment. The manager of the plant wants the candy bars to have a standard weight. Which graph shows less variation in candy-bar weight? *(Check to make sure that students are using observations about how clustered each graph is to support their claims.)*

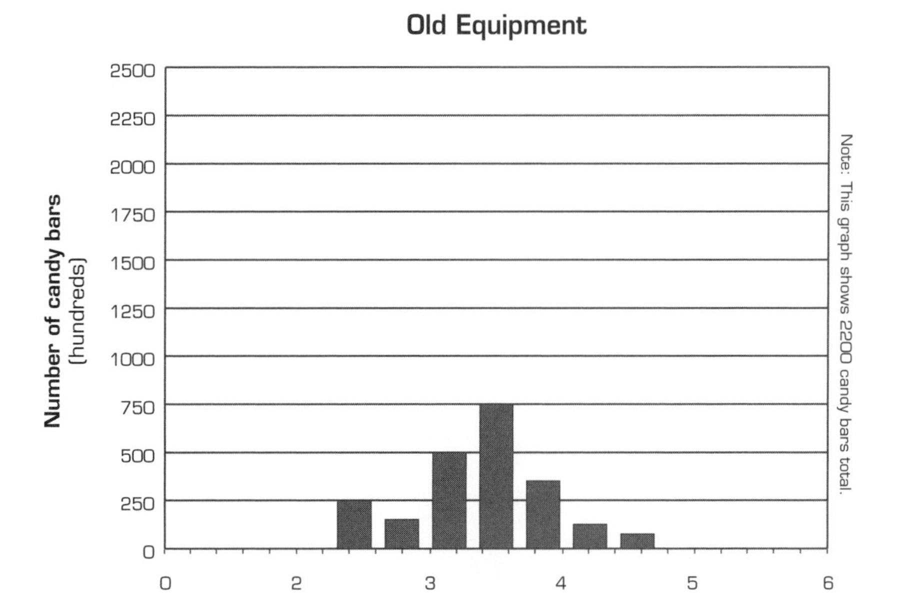

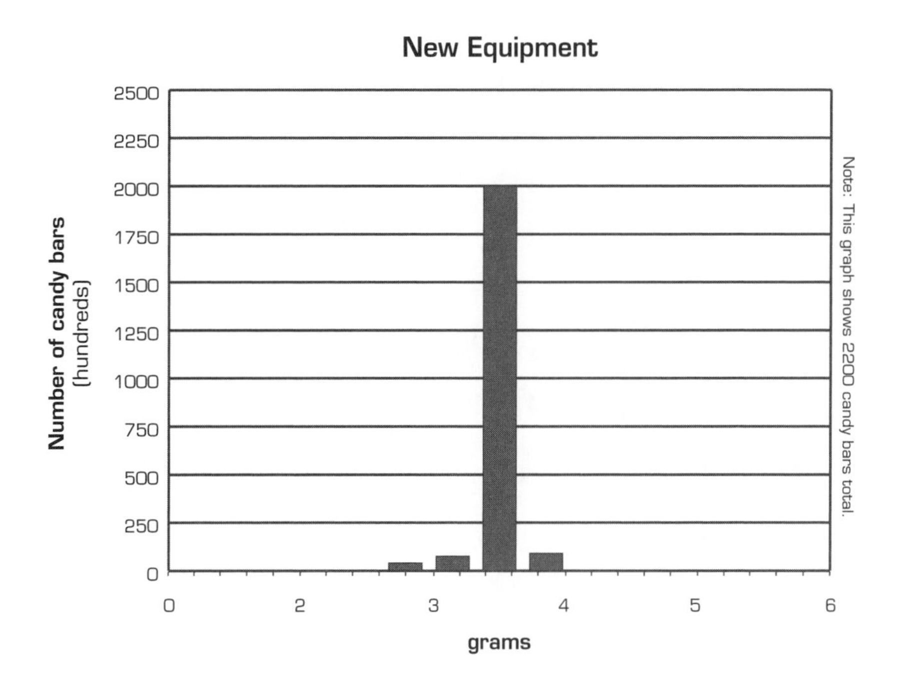

Preparation for 2.3

- **Science Content:** Graph what you think the class data would look like if the data were more reliable. Do you think there would be any spread in the results? If so, what might contribute to this spread? *(Students should draw a graph with clustered data. Students might realize that it will be difficult to eliminate all variation. When the class graphs the results of the investigation in* Section 2.3, *you can compare the actual results with what students expected.)*

NOTES

SECTION 2.2 IMPLEMENTATION

◄ $\frac{1}{2}$ *to 1 class period**

2.2 Investigate

Modeling Lava Flow I

5 min.

Students are introduced to the purpose and usefulness of models in science.

2.2 Investigate

Modeling Lava Flow I

Because you cannot use actual lava, you are going to make a **model** of lava flow. You will be using dish soap and a plastic plate to **simulate** lava flowing across a landscape. You might say that this model is not realistic because dish soap is not as hot as lava. That is true, but your model is not investigating the temperature of lava. You are looking at how fast lava flows, and dish soap flows very much like some types of lava. This makes the soap and plastic plate a good set of materials for modeling lava flow.

Models and simulations help scientists learn. The simulations you are doing will help you learn how to measure the rate at which lava flows. You will work with several other models and simulations in this Unit and throughout PBIS.

model: a representation of something in the world.

simulate: use a model to imitate or act out real-life situations.

Be a Scientist

Using Models and Simulations

A model is a representation of something in the world. One model that you know is a globe. The parts of the globe represent parts of Earth. Scientists use models to investigate things that are too difficult or too dangerous to examine in real life. The models they use are at a size that people can easily examine.

To use a model to investigate, the model needs to be similar to the real world in ways that are important for what the scientist is investigating. Sometimes, what you want to model is a situation or an event. To do this, you create a model that includes the things that are part of an event and then use that to act out a situation. These are called simulations. Simulations use a model to imitate, or act out, real-life situations. Simulations imitate, or act out, what happens in real life in a way that is similar to real life but lets you examine what is happening without causing any harm or danger. Scientists use simulations when what they want to study is too big or too small, too fast or too slow, or too dangerous to investigate directly.

Airline pilots often train in aircraft simulators. They climb into a machine that looks and feels like a real cockpit. This way they can make mistakes and learn from them without harming people and property.

DIG 31 DIGGING IN

META NOTES

The model used in this investigation (soap as lava and the plate as land) has many limitations. As with most models, the materials limit our ability to replicate how the system being modeled works. The soap and plate do provide an opportunity for students to design a procedure, collect data, and discuss how the model might provide information to answering the challenge. Students may feel some uncertainty about using a model to investigate how fast lava flows. One of the lessons they should draw from this investigation is that models and simulations are a useful way to answer questions, but as with any investigation, scientists need to be aware of the limits of the model.

△ Guide

Begin by asking students how experimenting with the model of dish soap and a plastic plate is different from experimenting with real lava on land. Students might say that the dish soap is not hot or dangerous, and a plastic plate does not have all the hills, depressions, and foliage of real land.

Ask students to describe some of their experiences with models. Some students may mention model planes or cars. These models are representations, in scale, of real things. Scientists use models when studying

*A class period is considered to be one 40 to 50 minute class.

something too big, too small, too fast, too slow, or too dangerous. Tell students that scientists would use the dish soap flowing over the plastic plate to model lava flowing over land. When scientists investigate a phenomenon using a model, they call it a simulation.

Emphasize that one limitation of a model is that it does not represent all the parts of a system. This model does not take into consideration the temperature of the lava, but it does consider the speed of the flow.

NOTES

Materials
- plastic plates
- stopwatch
- model lava (bottle of dish soap)
- paper towels
- *Lava Flow Data* page

Design Your Procedure

As a class, spend five minutes developing a procedure for measuring lava flow. Use the materials shown on the list.

Run Your Procedure

You will have 10 to 15 minutes to run your procedure. You have everything you need to model the flow of lava.

- Use seconds to measure the time it takes for the lava to flow across the plate. Round fractions of seconds to the nearest whole second (3.51 seconds to 4 seconds).
- Run at least five trials to show that the time you measured for the lava to flow is consistent.

You will need to record your data during this investigation. Remember, recording results allows scientists to accurately report their findings. Data help others understand a scientist's work. They also help other scientists do future investigations.

Record your results on a sheet of paper. Be prepared to share your results with your class.

Communicate Your Results

Share Your Data

The last time you communicated your work, each group presented in a *Solution Briefing*. This time you will do it differently. Each group will report to the class one result (amount of time in seconds that it took for the model lava to flow across the plate). As each group reads out their results, you will chart them on a **bar graph**. You will do this by placing an "X" on the graph for every data point on your *Lava Flow Data* page.

bar graph: type of graph that uses either vertical (up and down) bars or horizontal (across) bars to show data. Data can be in words or numbers.

By creating a bar graph, you will be able to demonstrate that your class can

- accurately determine how fast lava flows, and
- reproduce the same result over and over.

Remember, your class is a team of scientists, and towns will be counting on you to save lives. Your procedure for measuring lava flow needs to be accurate in order to keep everyone safe.

Design Your Procedure

5 min.

Students design a procedure to find how fast dish soap runs across the surface of a plastic plate.

META NOTES

The instructional goal of this investigation is to have students, in pairs, develop a procedure, perform the investigation, and see that the class data is inconsistent. Spoons and rulers could be useful to measure the soap and the distance it moves. If students have already mentioned these constraints as part of the previous discussion, supply rulers and spoons. Ideally, no rulers or measuring spoons would be used because students should have varied results to see they need standardized procedures.

△ Guide

Tell students they have approximately five minutes to develop a procedure they can use to answer the question: *How fast does the model lava (soap) flow (across the plate)?* Distribute the materials and guide the class's discussion of ideas for a procedure. If necessary, ask students questions about how they could measure lava to start the discussion.

Students probably will not identify all of the factors of this investigation at this point. If students do not discuss the angle they should tilt their plates, do not suggest an angle. Different groups may tilt their plates at different angles and measure very different times which will help students see the need for standardized procedures. When students compare the results of their first investigation and discuss the reasons for the variation, the angle of the plate will be identified as a potential constraint, and the class can decide on an angle for all pairs to use.

As students choose how to run their simulations, record their choices on the board or have students write them on a sheet of paper.

☐ Assess

As the class develops a procedure, assess students' contributions to the discussion.

Run Your Procedure

15 min.

Students run their investigations.

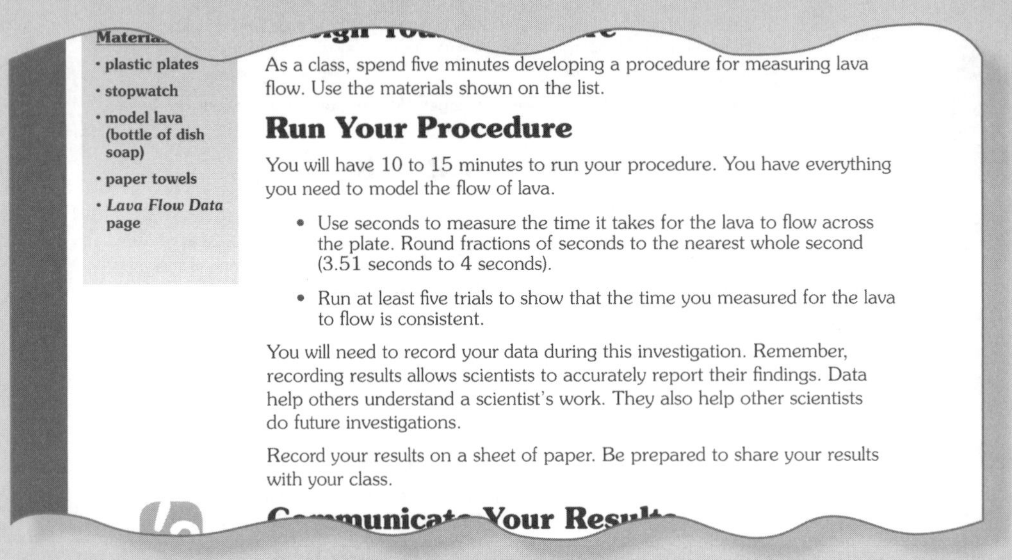

Materials
- plastic plates
- stopwatch
- model lava (bottle of dish soap)
- paper towels
- *Lava Flow Data* page

As a class, spend five minutes developing a procedure for measuring lava flow. Use the materials shown on the list.

Run Your Procedure

You will have 10 to 15 minutes to run your procedure. You have everything you need to model the flow of lava.

- Use seconds to measure the time it takes for the lava to flow across the plate. Round fractions of seconds to the nearest whole second (3.51 seconds to 4 seconds).
- Run at least five trials to show that the time you measured for the lava to flow is consistent.

You will need to record your data during this investigation. Remember, recording results allows scientists to accurately report their findings. Data help others understand a scientist's work. They also help other scientists do future investigations.

Record your results on a sheet of paper. Be prepared to share your results with your class.

⬡ Get Going

Once the class has developed procedures, give them about 15 minutes for their investigation. Emphasize the importance of keeping good records and having pairs record their results (one student can record the results while the other pours dish soap onto the plate). Remind students that good scientific practices include keeping good records.

Review the additional criteria in the reading. Remind students to use seconds to measure time and to run five trials.

△ Guide

As students run their procedures, check how they carry them out. Sometimes they will do things differently from what they have written (e.g., they will hold the plate at a different angle, they will use a different quantity of dish soap, or they will time the flow of the soap differently). Do not correct students, but note inconsistencies within and across pairs. You will be helping students identify what led to the large spread in the results later. Encourage and model the way students will begin evaluating their results. Probe students by asking about the variability in their results. Ask pairs with wide results across trials, "Why do you think your results are so varied?" Finally, note the highest and lowest times students are getting for their results. This will give you an idea how to label the columns in the bar graph for communicating results.

Later, when students are trying to determine why their results were inconsistent, you can identify specific procedures that were inconsistent or ways pairs did not follow their procedures to help students identify the causes of the varied results.

Some pairs might notice that their results are different from the results of another group. Let them know they will share their results and compare them with the rest of the class.

□ Assess

Sometimes students have difficulty focusing on the task to start. Remind them to move quickly through the task. If a pair has not started after about five minutes, check to make sure they understand the task and encourage them to focus on the task.

NOTES

META NOTES

Without guidance, students will select different angles to tilt their plates. The tilt is a very important variable. Different angles yield flow times between five to 10 minutes. The total data collection time should be no more than 15 minutes. Optimally you want pairs to have a minimum of five trials, however, if a pair is measuring times of five to 10 minutes they will not be able to do five trials. This is okay as it will provide more variation in the procedures and times measured and will reinforce the need for standard procedures and repeatable results. If your class is large, each pair should run five trials; if your class is small, up to 10 trials. This ensures that the class will generate enough data to see inconsistencies. Approximately 70 trials will provide enough data and show the inconsistencies.

META NOTES

It is important to begin modeling scientific language for students. In this case, the language is about procedures and data gathering. Using words like "varied" or "range" in the correct scientific way, is appropriate and recommended.

Communicate Your Results

10 min.

Students collect and graph the class's data.

Record your results on a sheet of paper. Be prepared to share your results with your class.

Communicate Your Results

Share Your Data

The last time you communicated your work, each group presented in a *Solution Briefing*. This time you will do it differently. Each group will report to the class one result (amount of time in seconds that it took for the model lava to flow across the plate). As each group reads out their results, you will chart them on a **bar graph**. You will do this by placing an "X" on the graph for every data point on your *Lava Flow Data* page.

bar graph: type of graph that uses either vertical (up and down) bars or horizontal (across) bars to show data. Data can be in words or numbers.

By creating a bar graph, you will be able to demonstrate that your class can

- accurately determine how fast lava flows, and
- reproduce the same result over and over.

Remember, your class is a team of scientists, and towns will be counting on you to save lives. Your procedure for measuring lava flow needs to be accurate in order to keep everyone safe.

DIG 32

Project-Based Inquiry Science

△ Guide

When all pairs have finished running their procedures, let students know you will graph their results, and they will use the graph to determine the time it takes the model lava or dish soap to flow across the plate. Explain that you will mark X's on the class graph for student trials and that the height of the column of X's at any number along the horizontal axis shows how many times that number of seconds was measured during the investigation.

TEACHER TALK

"One way we can visualize all the data you have collected is to create a graph. I've labeled the *x*-axis with the number of seconds that it might have taken the lava to flow down the plate. Now I'm going to put an X on the graph for each measurement. If one pair got 30 seconds, I'll put one X over the 30-second mark on the graph. This is called a bar graph."

Ask each pair to report their data and plot the data on the graph on a transparency. Make sure that students plot the data on their own graphs. Provide additional time at the end of the data collection for students to check their graphs against the class graph and make sure they have accurately copied the data.

Determine the highest and lowest values students measured so that you will know how to label the columns in your bar graph. All data should be recorded together on one bar graph.

Review selected data with the students to assess their understanding of the graph. For example, make sure they know that if there are eight X's in a column, it means eight trials resulted in that measurement.

Remind students of the question they are trying to answer, *How fast does the model lava flow?* Ask them to derive and answer based on their data. Since the data are wide spread, it will be difficult to derive a conclusive answer from the graph. You might ask if this seems right or reasonable. Would students expect the dish soap to take a different amount of time to flow across the plate each time? They should begin asking questions about why the data cannot provide a conclusive answer.

NOTES

META NOTES

When students are confused by the inconclusive answer, they may assert alternative answers. Students may pick the number with the highest frequency (the mode). Students may pick the middle value (the median), in which half the data points are above the median and half the data points are below. They may also confuse this with the mean or average.

It is likely that students will suggest finding the average (the mean) by adding the numbers and dividing by the number of trials. The mean is the best value for multiple measurements of the same thing because it gives the value most likely to have the least amount of error. All measurements have some error, but the mean does not answer the question in this case because the range of the data is large and there are probably several outliers.

Analyze Your Data

15 min.

Students see that their data are inconsistent and inconclusive from their graphs and discuss finding the reason their data are inconsistent.

Analyze Your Data

Look at the bar graph. Work as a class to answer the following questions. Discuss how your answers may help you better achieve the *Lava Flow Challenge*. Have your written procedure available as you answer the questions.

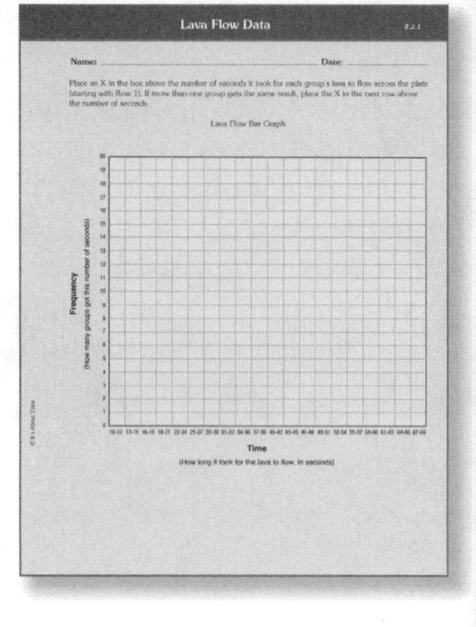

1. Did your group have any difficulties (mistakes, spills, etc.) while running the procedures? Describe each one.
2. How similar are the results of different groups?
3. What did the distribution, or spread, of data on the bar graph look like? What do you think this says about how reliable the class's data are? Do you think the town council will trust your results?
4. Why do you think there are differences in the data from different groups?
5. What could the class do to get more consistent results?
6. Do you think the company that hired you to develop a procedure measuring lava flow for towns near active volcanoes will trust your results?

What's the Point?

Most likely, the distribution of data on your class bar graph was spread very widely. This indicates that the results are not reliable. There may be many reasons why your results varied so much. However, one of the main reasons is that different groups used different procedures.

Scientists also face this problem. To confirm the results of other scientists, they run investigations again, following the exact same steps as the original scientist. If a scientist did not provide precise procedures, results cannot be accurately duplicated.

△ Guide

Guide students in analyzing their graphed data. Focus students' attention on their procedures as they consider why their data are inconsistent.

⬡ Get Going

Tell students to use their copy of the class graph and their own procedures as they think about each of the questions in their text. Give pairs about five minutes for this discussion.

If students are already asking each other these questions and making these suggestions, you can move ahead to the class discussion.

△ Guide

Monitor students' discussions. Assist student pairs in analyzing their graphed data. Focus students' attention on their procedures as they consider why there are inconsistencies in their data. If students are not beginning to recognize that their data were inconsistent because their procedures were inconsistent, ask them to tell you about how they poured the dish soap or how they measured the time, or how they tilted their plate.

□ Assess

As pairs discuss the questions, listen for their ideas about their procedures and the effect on their results. These ideas will be developed further in the class discussion.

△ Guide

Lead a class discussion to explore the effects of using procedures that are not standardized.

Use the questions in the student text as a guide to get the class thinking about how important it is to have standardized procedures to get consistent and trustworthy results.

TEACHER TALK

"Let's talk about why the range of our results is so large. Why do you think this happened? The questions in the student text will help us answer this."

1. Focus students' thinking on pairs' procedures. You can ask a pair or two to share how they set up their plate and how they poured the soap. Students will notice differences such as how they tilted the plate or when they started and stopped the watch. Students should describe specifically what went wrong.
2. Focus students' thinking on how pairs ran their procedures. Have two pairs with similar results dicuss how they ran their procedures.
3. Guide students' attention to the data and how wide the distribution is. Students' answers should show understanding that the distribution of data is wide because different procedures and materials were used. They should also realize that a wide spread in data suggests that the results are untrustworthy.
4. Consider data across pairs. Students should note the difference in procedures and how this can lead to different results.

META NOTES

Students will likely bring up accidents at this point, but listen for suggestions about having everybody know what to do better—using more precise and standardized procedures.

5. Students should see that to have trustworthy results they will need to ensure that they carefully follow the same procedures as everyone else. Listen for students' ideas about making sure everyone is careful running their procedures and measuring time the same way for each trial. You may guide students by asking what would indicate trustworthy results.

6. This question is meant to help students gain a concrete understanding of the importance of consistent results and reliability. Students should support their answers with reasons.

TEACHER TALK

"Obviously, it is difficult to get consistent data with inconsistent results. It would be difficult to use the data we have collect to answer the question about lava flow rates. If we still want to answer the question what should our next steps be?"

◇ Evaluate

Make sure that students understand for their results to be considered trustworthy they have to have repeatable results. Repeatable results, across time and different scientists, are possible if everyone runs a standard procedure.

NOTES

Assessment Options

Targeted Concepts, Skills, and Nature of Science	How do I know if students got it?
Graphs are an effective way to communicate results of a scientific investigation.	**ASK:** How did the graph help you understand the data? **LISTEN:** Students should recognize variation in the results they graphed. From this, they can conclude that their results were not reliable.
Scientists must keep clear, accurate, and descriptive records of what they do so that they can share their work with others and consider what they did, why they did it, and what they want to do next.	**ASK:** Why is keeping a record of what you have done important? **LISTEN:** Students should describe the importance of keeping good records so that they can accurately describe what they did, why they did it, and what they want to do next.
Scientific investigations and measurements are considered reliable if the results are repeatable by other scientists using the same procedures.	**ASK:** Does it matter if you record at what angle you held your plate? **LISTEN:** Students should see that keeping records of their investigation findings help them and others replicate the procedure.
Scientists use models to simulate processes that happen too fast, too slow, on a scale that cannot be observed directly (either too small or too large), or that are too dangerous.	**ASK:** Why do scientists use models? **LISTEN:** Students should describe models as useful when scientists are trying to study a process that cannot be observed easily or safely.

NOTES

Teacher Reflection Questions

- What difficulties did students have making the connection between the wide variation in data and the need for standard procedures and materials? How did you assist them in making this connection? What else could you do?
- Standardized procedures are not given to students right away so that students will recognize the need for them through experience. What discomforts did your students show from not being given a set of procedures right away? What discomforts did you have?
- What worked well and what could you do better during the collection and analysis of the class data?

NOTES

SECTION 2.3 INTRODUCTION

2.3 Plan Your Investigation

Getting to a Better Procedure

◄ ***1 class period***

A class period is considered to be one 40 to 50 minute class.

Overview

Students identify factors in their procedures that led to inconsistent results, and then design a more precise class procedure to control these factors. They design a class procedure that is detailed and replicable, controlling each factor they identified. Students reflect on their new procedure by comparing it to the original procedure and they see how their ability to plan an investigation has improved.

Targeted Concepts, Skills, and Nature of Science	Performance Expectations
Identifying factors that could affect the results of an investigation is an important part of planning scientific research.	Students identify factors that can affect the results of their investigation.
Scientific investigations and measurements are considered reliable if the results are repeatable by other scientists using the same procedures.	Students describe how investigations and measurements are considered reliable if they can be repeated by following the same procedures. Supporting reasons should include observations from the previous section, such as that when groups used different results, they all came up with different answers and they couldn't decide which answer was correct.
Scientists often work together and then share their findings. Sharing findings makes new information available and helps scientists refine their ideas and build on others' ideas. When another person's or group's idea is used, credit needs to be given.	Students consult their peers in planning, and they share their results with their peers.

Materials	
1 per class	Media for creating class procedures (e.g., butcher block paper, overhead) Class list of criteria and constraints

Homework Options

Reflection

- **Science Content:** Scientists in City A spent two hours one evening catching fish in a stream using an experimental new bait and caught five fish. Scientists in City B spent three hours one morning catching fish in a lake using the same bait and caught two fish. They could not reach any conclusion about how many fish they can catch using the bait. What are the factors that led to the different results? What are some practical ways the scientists could reduce the variation? *(Check to see if students' answers address the differences between the studies. In this case, a difference in settings—stream vs. lake—and a difference in time.)*
- **Science Process:** Think about the criteria and constraints you listed for your first investigation with the plate and dish soap. Identify some additional criteria and constraints that will guide your next investigation with the plates and dish soap. *(Students' answers may include: criteria—a more specific set of procedures, constraints—number of trials.)*
- **Nature of Science:** Describe the reasons for creating a clear, standard procedure. *(Students' answers should point out the need for other researchers to be able to run the same procedure and get similar results. Otherwise, the results are not considered reliable.)*

Preparation for 2.4

- **Science Content:** If everyone in class runs the class procedure in exactly the same way, what do you expect the bar graph of the results will look like? Draw a picture to illustrate your answer. *(Students' answers should say something about the data being more clustered. The picture should depict a bar graph with tightly clustered data.)*

SECTION 2.3 IMPLEMENTATION

2.3 Plan Your Investigation

Getting to a Better Procedure

Your class probably did not agree on how fast the dish soap flows. Your line plot may have shown that your class cannot produce reliable results. You will now see if you can find a way to make the results more consistent across groups.

Think about what went wrong. You were all trying to answer the same question. You all measured the flow of dish soap across a plastic plate. You all used the same unit of measurement. You all had the same materials. But every group used a slightly different procedure. You all collected data in different ways. No wonder the results were so varied.

repeatable: when someone follows the reported procedure, they get similar results.

replicate: to run a procedure again and get the same results.

soil: the loose top layer of Earth's surface, able to support life, consisting of rock and mineral particles mixed with organic matter, air, and soil.

Be a Scientist

Designing Good Procedures

Scientists only trust results that are **repeatable** by other scientists. In order for other scientists to **replicate** the results of an investigation, the procedures must be reported very precisely. Then someone else can run the procedure again and get the same results.

Making Procedures Repeatable

For example, suppose you wanted to investigate the effect of a fertilizer on the growth of plants. You would need to keep many other factors the same:

- **soil** type,
- time spent in sunlight each day,
- amount of water, and
- type of plant.

Think about one factor, water.

You would need to make sure that each group of plants got the same amount of water. They would need to be watered the same number of times. Also, they would need to be watered in the same way.
You would need to follow these rules every single time you watered each plant.

◀ *1 class period*

2.3 Plan Your Investigation

Getting to a Better Procedure

15 min.

Students design a standard class procedure to get a smaller distribution of data.

○ Engage

Now that students have seen evidence of their inconsistent data in their bar graphs, and have thought about why their results were inconsistent, ask if they can design a procedure that would get better results. Display the class graph during this discussion.

TEACHER TALK

"Look at our class graph again. There are a lot of different results. You've thought about what made your results different. Why do you think the results are so varied? Scientists are interested in procedures that can be repeated by others and provide consistent data. What can you do to make a procedure that would help your class results be less varied?"

△ Guide

Begin by reminding students of the differences in procedures and materials they identified and some of students' ideas for getting more reliable results.

Lead the class in identifying all the things they will need to pay attention to when revising their procedure.

TEACHER TALK

"Let's think about all the factors that need to be the same from trial to trial—things we need to think about in our new procedure.

What are some of the things you did differently from other groups? What are some of the things you should have done the same?"

As students identify factors in their investigations and ways their procedures were not standardized or specific, record the factors so that students can refer to this list when the class creates a standard procedure in the next step.

Many of the factors identified should involve inconsistencies with measurement. Discuss how students should use the same measurement. Emphasize that a standard procedure will need to describe how each of these factors will be measured and controlled.

Students should begin to see that the tilt of the plate changes the speed of the flow. This is one very important factor. If students have not discussed this, remind them how much the plate tilts affects how much time the dish soap takes to cross the plate.

The volume of the dish soap will also affect how fast it flows. Have the class agree on an amount of lava to put on the plate and how they will measure it. Tell students that measuring spoons are available.

Connect students' work to how scientists design good procedures by reviewing *Designing Good Procedures* in the *Be a Scientist* textbox. Emphasize that scientists trust experimental results that are repeatable by other scientists, and so they also need to know how to replicate their procedures. For this reason, procedures must be detailed and descriptive.

Measurement and Precision

It is also important to make the same measurement each time. You could count the number of leaves on each plant, or you could measure the height of each plant. Or, you could do both.

The tools you use can often affect measurement. You have limits to what you can see when you make a measurement. Be sure to consider how accurate the tools you use are.

Here is a checklist that you can use to make sure your measurements are consistent:

- Measure from the same point.
- Measure with the same units.
- Repeat **trials** for more **precision**.
- Start fresh. Do not compare data from before you make a change to a procedure to data collected after you make a change.
- Measure under the same conditions.

Errors and Measurement Constraints

Some groups in the class probably reported mistakes, or errors, they made as they were running their procedures. There are two kinds of errors to think about when you design procedures. Some mistakes are avoidable and others are not. For example, forgetting to start the stopwatch when the dish soap begins to flow down the plate is an avoidable mistake. These are the easiest kinds of errors to fix. You usually know when you have made this type of mistake and you can easily run the procedure again.

But some errors are unavoidable. Every measurement has error that is impossible to avoid. If you have a ruler that measures in millimeters, you sometimes cannot tell if something measures half a millimeter or a quarter-millimeter. Even if you got a better ruler, there would still be a point at which you would have to estimate. Your results can only be as accurate as your measuring tools. Because scientists have constraints on their tools, they will always make measurement errors that are unavoidable. Scientists usually keep track of possible measurement errors when they report on an investigation. This way, other scientists can judge how much to trust the results.

trial: one time through a procedure.

precision: how close together the measured values are.

Emphasize measurement, precision, and errors. Let students know the importance of repeating trials so that they can be assured their measurement is reliable. It is also important for students to realize that however precise their measurements are, there will always be some unavoidable error which is at the very minimum due to the constraints of their measuring device.

Revise Your Procedure

15 min.

The class decides upon a standard procedure for the investigation.

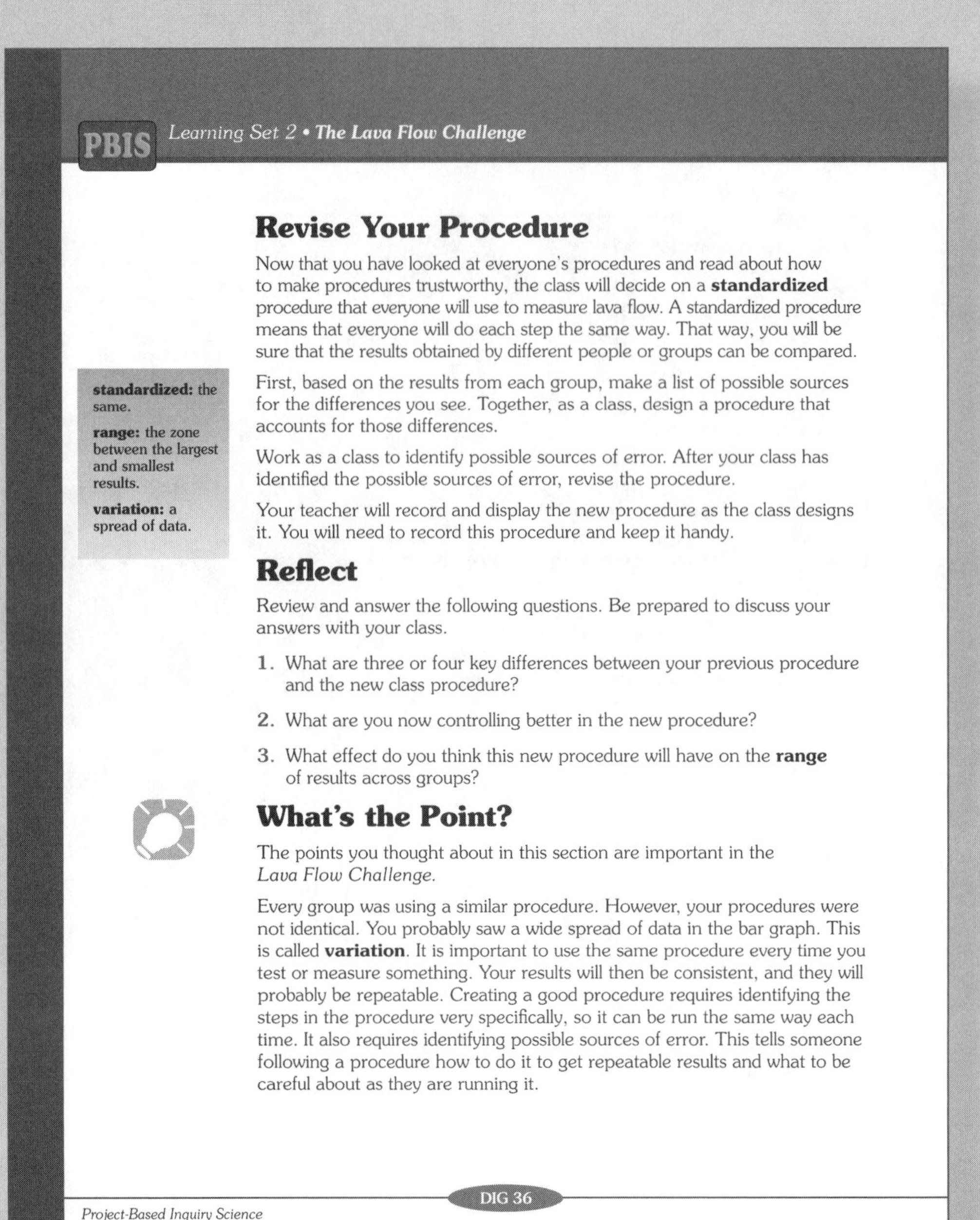

Revise Your Procedure

Now that you have looked at everyone's procedures and read about how to make procedures trustworthy, the class will decide on a **standardized** procedure that everyone will use to measure lava flow. A standardized procedure means that everyone will do each step the same way. That way, you will be sure that the results obtained by different people or groups can be compared.

standardized: the same.

range: the zone between the largest and smallest results.

variation: a spread of data.

First, based on the results from each group, make a list of possible sources for the differences you see. Together, as a class, design a procedure that accounts for those differences.

Work as a class to identify possible sources of error. After your class has identified the possible sources of error, revise the procedure.

Your teacher will record and display the new procedure as the class designs it. You will need to record this procedure and keep it handy.

Reflect

Review and answer the following questions. Be prepared to discuss your answers with your class.

1. What are three or four key differences between your previous procedure and the new class procedure?
2. What are you now controlling better in the new procedure?
3. What effect do you think this new procedure will have on the **range** of results across groups?

What's the Point?

The points you thought about in this section are important in the *Lava Flow Challenge.*

Every group was using a similar procedure. However, your procedures were not identical. You probably saw a wide spread of data in the bar graph. This is called **variation**. It is important to use the same procedure every time you test or measure something. Your results will then be consistent, and they will probably be repeatable. Creating a good procedure requires identifying the steps in the procedure very specifically, so it can be run the same way each time. It also requires identifying possible sources of error. This tells someone following a procedure how to do it to get repeatable results and what to be careful about as they are running it.

META NOTES

Modeling scientific thinking is critical. Students do not have the experience to know what scientific thinking sounds like. By using a think-aloud method in which you verbalize your thoughts, you give students a model for what scientific thinking might sound like. Modeling is particularly helpful since this is the first time students have designed a procedure.

△ Guide

Lead the class in designing a new procedure that will give a more reliable answer. Students should be reminded to account for all factors and identify how the factors will be the same. Remind students that the goal is to develop a specific class procedure that every group can reproduce. By using the same procedures, students should get similar results.

One way to lead students in designing a new procedure is to model your thinking as you focus on one factor at a time. You might want to talk about the amount of dish soap first.

TEACHER TALK

"Let's start with the dish soap. We could use different amounts of dish soap, and we could drop or place the dish soap on the plate differently. How much dish soap should we use? Some groups used more soap than others. What happened when groups used a lot of soap? What about when they used a small amount? Did they put the dish soap on the plate when it was tilted or flat? How should we put the dish soap on the plate and start the experiment?"

META NOTES

Designing a procedure is a good opportunity to assess students' ideas about controlling factors. In this case, students should be thinking about how to measure all the things that might affect how fast the soap flows across the plate.

As you think through the procedure with students, record the steps so that students will be able to evaluate the procedure and use it to run their investigations.

☐ Assess

As you develop the procedure, listen to students' ideas about how to measure consistently and precisely and how to control factors in investigations.

△ Guide

When the procedure is complete, go through the entire procedure with the class. Clarify any steps that might be ambiguous.

Reflect

10 min.

Students compare the old and new procedures.

△ Guide and Assess

Use the *Reflect* questions to assess students' understanding of what they have done and why. You might have students write their answers to the *Reflect* questions, and then lead a discussion of students' responses. Alternatively, you might want to lead a discussion of the questions without having students answer them first.

1. Students' responses should show an understanding of what was important about their revisions. The key changes will probably be in the steps where there were differences between groups' procedures.
2. The factors in students' lists should correspond to the key revisions to the procedure. In this discussion, emphasize changes made in the steps where there were differences in the way groups carried out their procedures and where factors were not specified. Students may not understand the word controlling. Consider giving an example such as controlling the tilt of the plate or the volume of dish soap used.
3. The new procedure should reduce the range of the results. This is the first time the word range is used. Emphasize to students that range is the spread in the data.

Assessment Options

Targeted Concepts, Skills, and Nature of Science	How do I know if students got it?
Identifying factors that could affect the results of an investigation is an important part of planning scientific research.	**ASK:** What are some of the reasons for inconsistent results from an investigation? **LISTEN:** Students should recount things that happened during their investigations that led to variation. **ASK:** What should the class graph look like when the investigation results are more consistent? **LISTEN:** Students should expect the class graph to be more clustered.
Scientific investigations and measurements are considered reliable if the results are repeatable by other scientists using the same procedures.	**ASK:** How can you verify that another researcher's results are reliable? **LISTEN:** Students should know that if results are reliable, a researcher should be able get similar results by following the same procedure. **ASK:** How can changing your procedure from trial to trial affect your results? Give an example. **LISTEN:** Students should link changing procedures to varied results and use examples from their investigation.
Scientists often work together and then share their findings. Sharing findings makes new information available and helps scientists refine their ideas and build on others' ideas. When another person's or group's idea is used, credit needs to be given.	**ASK:** Why is it important for researchers to have carefully controlled procedures? **LISTEN:** Students should recognize that a researcher needs to have results verified by peers, and that a researcher needs to establish procedures that peers can replicate.

Teacher Reflection Questions

- What evidence do you have that students understand the need for a clear procedure that can be replicated and the need for repeatable results?
- What types of attitudes toward successes and mistakes have you observed in your students so far? Did some students compete for the best results in either investigation? What types of changes in these attitudes have you observed so far in this Unit?
- What did you do to model appropriate language during discussions? What ideas do you have for next time?

NOTES

SECTION 2.4 INTRODUCTION

2.4 Investigate

Modeling Lava Flow II

1 class period ▶

A class period is considered to be one 40 to 50 minute class.

Overview

Students run their revised lava flow procedure and collect data, which they share on a class bar graph. Using the spread of the data to evaluate their new procedure, the class identifies ways that the procedure is still not precise enough. Students learn about different types of lava flow and revise their procedures and share their data. This time, their results should be more consistent, providing evidence that a precise, standard procedure ensures consistent results.

Targeted Concepts, Skills, and Nature of Science	Performance Expectations
Scientists often work together and then share their findings. Sharing findings makes new information available and helps scientists refine their ideas and build on others' ideas. When another person's or group's idea is used, credit needs to be given.	Students should consult their peers in planning and share their results with their peers.
Scientific investigations and measurements are considered reliable if the results are repeatable by other scientists using the same procedures.	Students should follow the class's standard procedure, analyze the class's data, and describe, based on their experiences, why scientists consider results reliable if they are repeatable.
When volcanoes erupt, magma reaches Earth's surface and is called lava. There are many different types of lava.	Students should be able to describe that there are different types of lava and that these may flow at different rates.

Homework Options

Reflection

- **Science Content:** Based on the results of your last investigation, what could you say about the results other researchers using your procedure would get? Why? *(Students should be linking the reliability or unreliability of the results to their usefulness for saying what results other researchers might get.)*
- **Science Content:** If you wanted to model and measure the flow rates of different types of lava (basalt, andesite, dacite, and rhyolite), what changes would you need to make to your lava model? How would you change the materials to better model different lavas? *(Students should describe that the model would have to include new test materials. The materials would need to have flow rates that mimicked the different types of lava. Slow-flowing lava might best be modeled using honey or molasses, while fast-flowing lava might best be represented by water or milk.)*
- **Science Content:** You have a set of measurements of the time it takes a pendulum to swing back and forth 10 times. The results are as follows:

Trial	Time(s)
1	14.7
2	15.1
3	15.3
4	21.3
5	14.9

Make a bar graph for these results. Do you think all the trial values are reliable? Why or why not? Can you reliably say how long it takes the pendulum to swing back and forth 10 times? If so, how long? *(Students should identify the outlier at trial 4, and if they excluded the trial from analysis.)*

NOTES

SECTION 2.4 IMPLEMENTATION

◀ *1 class period**

2.4 Investigate

Modeling Lava Flow II

Run Your New Procedure

Now that you have a new procedure, your class should be able to produce more reliable results. Your class will soon collect another set of data and produce a new bar graph. As a class, update the criteria and constraints of the challenge if you need to.

Follow your new procedure. Use the materials listed. Obtain results for 5 to 10 trials.

Record your results on the same sheet of paper you used to record your procedure. Be prepared to share your results with your class and teacher. You will have 10 to 15 minutes to perform your procedure and collect your data.

Materials
- plastic plates
- stopwatch
- model lava (bottle of dish soap)
- paper towels

Communicate Your Results

Share Your Data

Use another sheet of graph paper to make a bar graph from the new data.

As before, each group will read aloud their results. Everyone will plot them on the graph paper. Each group should report any problems they had running the procedure (e.g., mistakes, spills).

Analyze Your Data

After your class creates the second bar graph, answer the following questions together:

1. How do the results from this investigation compare to the ones from your first set of trials?
2. Did all groups get results similar to yours?
3. Do you trust these results more? Why or why not?

DIG 37 DIGGING IN

2.4 Investigate

Modeling Lava Flow II

Run Your New Procedure

10 min.

Students run an investigation with the new procedure that the class designed in Section 2.3.

△ Guide

With the class, update the list of criteria and constraints from *Section 2.1.* The criteria and constraints should include specific requirements for reliable results, such as: groups need to carefully follow the procedure the class decided upon and they need to ensure that they are all using the same materials.

*A class period is considered to be one 40 to 50 minute class.

Get Going

Distribute the materials or have students get them from a materials station and ask students to begin their investigations, emphasizing that pairs should record their results on the same sheet of paper where they wrote their procedure. One student can record results while the other puts the soap on the plate, as they did in their first investigation. Make sure students know how long they have (10 or 15 minutes) and how many times to repeat the procedure. Pairs should run five trials if your class is large and 10 if your class is small. This ensures that the class will generate enough data (about 70 total trials total for the class).

Assess

While students are running their procedures, you can observe how closely groups are following the class's standard procedure. If any group is doing something differently from the rest of the class, note the difference as something to discuss later. Look at the data to get an idea of how consistent they are. Listen for students' ideas about carefully following a procedure or recording data.

Communicate Your Results

Share Your Data

10 min.

Create a class graph of students' data after all pairs have finished running their investigations.

Communicate Your Results

Share Your Data

Use another sheet of graph paper to make a bar graph from the new data.

As before, each group will read aloud their results. Everyone will plot them on the graph paper. Each group should report any problems they had running the procedure (e.g., mistakes, spills).

Guide

Make sure that all students have graph paper. Briefly review how graphs help scientists share and analyze data, especially when there are many data points and they are looking for trends. Using a clean transparency, ask each pair how many seconds they recorded for each trial. For each result, put an X on the graph. Remind students to record the results of the class on their own graphs.

TEACHER TALK

"We're going to plot your data on a graph just like we did last time. Remember, each of the columns on the bar graph represents a number of seconds that it might take for the dish soap to flow down the plate. If a group got 40 seconds, I'll put an X over the 40-second mark on the graph. You should also do this on your graph. When we are finished, I will allow some time to check your graph against the class graph."

Analyze Your Data

After your class creates the second bar graph, answer the following questions together:

1. How do the results from this investigation compare to the ones from your first set of trials?
2. Did all groups get results similar to yours?
3. Do you trust these results more? Why or why not?

Analyze Your Data

35 min.

Students evaluate their data, looking for evidence that their procedure is refined enough to limit the range of the results. If the data are still very scattered, students revise their procedure again.

△ Guide

Lead students to examine their graphed data and critique their new procedure. Begin by looking at the spread of data and the precision of the procedures. The goal is to evaluate whether the procedure must be refined and repeated once again to get a reliable result. The answers to the first two questions in the student text will provide evidence for discussion about the trustworthiness of this new data (Question 3). Use these points to guide the discussion:

1. If students have effectively redesigned their procedure, they should find that the results of the second investigation are more consistent. Consistent data should have less range and should be more clustered on the bar graph.
2. It is likely that the class will have a few outliers. Students may be able to connect these with mistakes or accidents of some kind that happened during the investigation. Encourage students to share the circumstances around the outliers. This is a good opportunity to discuss what should happen to the numbers that are obviously faulty. Ask if students should disregard them, repeat them, or average them in. Each decision has different implications for the investigation. You might also discuss the fact that no measurement is ever exact. There is always some range of values that the measurement may fall within.
3. Students may say results that are more consistent can be repeated, and results that can be repeated are more reliable or trustworthy. Move the discussion toward how scientists determine if a result is trustworthy. Scientists call results trustworthy or reliable if the trends in the results found by others following the same procedure are the same.

As students evaluate the results of the investigation, discuss whether the results are consistent enough to determine a speed for the lava flow.

META NOTES

When an experimental data point is an outlier, it is usually due to a blunder or mistake while running the investigation. If this is the case, then the data point should be noted as due to a mistake, and should not be included in the data analysis. If possible, that investigation should be run again to collect another data point. If there is no evidence that the outlier is from a blunder, then it should be included in the data analysis.

Revise Your Procedure

< 10 min.

Students revise their procedures for a third iteration.

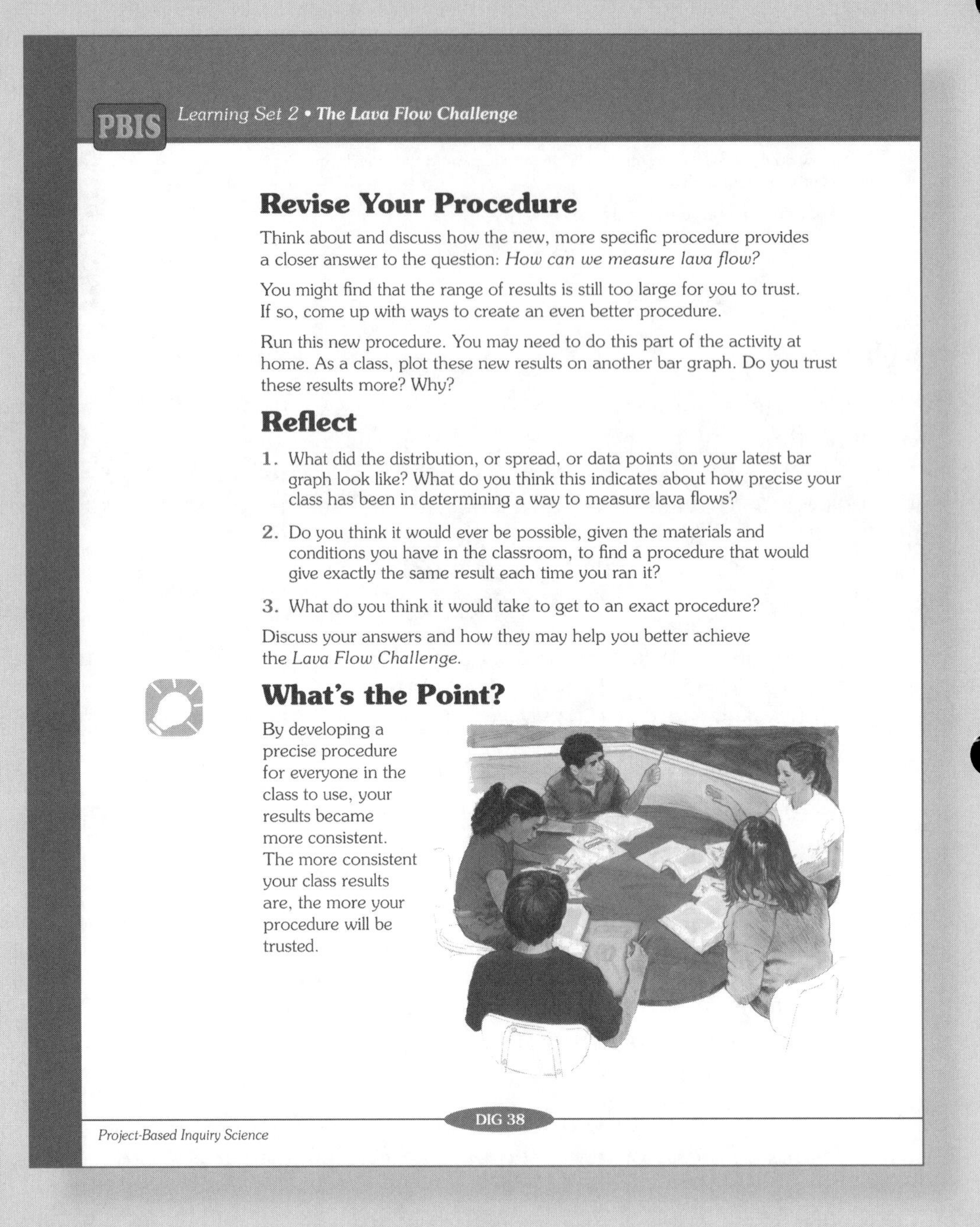

Revise Your Procedure

Think about and discuss how the new, more specific procedure provides a closer answer to the question: *How can we measure lava flow?*

You might find that the range of results is still too large for you to trust. If so, come up with ways to create an even better procedure.

Run this new procedure. You may need to do this part of the activity at home. As a class, plot these new results on another bar graph. Do you trust these results more? Why?

Reflect

1. What did the distribution, or spread, or data points on your latest bar graph look like? What do you think this indicates about how precise your class has been in determining a way to measure lava flows?
2. Do you think it would ever be possible, given the materials and conditions you have in the classroom, to find a procedure that would give exactly the same result each time you ran it?
3. What do you think it would take to get to an exact procedure?

Discuss your answers and how they may help you better achieve the *Lava Flow Challenge.*

What's the Point?

By developing a precise procedure for everyone in the class to use, your results became more consistent. The more consistent your class results are, the more your procedure will be trusted.

△ Guide

If you decided the class should revise their procedures again, guide students to think about the problematic differences and parts they determined earlier. Require them to be more specific as you lead the class in designing a new procedure. Since this is possibly the third time students have worked on this procedure, focus on refining or adjusting the current procedure rather than starting from scratch.

TEACHER TALK

"Scientists use iteration to improve their scientific procedures. You have revised your procedures and seen what happened to your results. Do we need to "tighten up" the procedures to make the results more accurate? What did you notice about the procedures and results of others? What slight differences did some groups have in their procedures that would make their results differ?"

META NOTES

When you monitor groups you may notice slight differences in some groups' implementation of the procedures. Guide students to recognize the source of inconsistent results.

As students specify steps, record the new procedure. Once the class has revised their procedure, have student pairs run their investigations. Graph the new data on a bar graph with the class, and help them analyze the data and evaluate the results.

△ Guide

To get students thinking about the importance of precise and standardized procedures, have them look at the graphed data from each iteration of their investigation. If students have three graphs, first focus on comparing the graphs and procedures. If they have only two, then they already made the comparison.

Focus students' attention on the spread of the data and remind them of the changes they made in their procedures that led to the improvements in the reliability of the data.

Reflect

15 min.

Now that students have two or three iterations of their investigation, they participate in a class discussion comparing the procedures and resulting graphs across the investigations.

TEACHER TALK

"How has the spread of the data changed? What changes in your procedures could have changed the spread of the data?"

Listen for the following in students' answers and guide the class toward these responses if necessary. Help students make a connection between their responses and how they will help them better achieve the challenge.

1. Students' responses should demonstrate an evaluation of the data, and an understanding of what the distribution of a data set indicates. If students say data are clustered tightly around a number, they should use this as evidence that they have used precise procedures. If students say the data are still scattered, they should use this as evidence that their procedure still is not precise enough. In order to achieve the challenge they need to have results that are trustworthy. The results are more trustworthy when their data are more clustered.

META NOTES

The average speed can be calculated by taking the total distance traveled divided by the total time. In this case, the distance would be that of the dish soap across the plate (which should be measured with a ruler) and the time would be the average time the class obtained, whose reliability depends on how clustered the data are.

META NOTES

Groups may still be straying from the standard procedure, causing varied results. If you observed any groups straying, you can lead students to this source of variation. If groups are following the standard procedure and still finding different ways of doing things, then the procedure needs to be more precise, and you can lead them to this after the reading.

TEACHER TALK

"How would you describe the range or spread of the data on the each graph? On which graph do you see the data most spread? Why do you think there is less range on this graph (present the second or third graph)?

What were the most important changes you made to your procedure to improve data collection? What changes do you think made data collection more difficult and perhaps less accurate? Is the new procedure more or less precise than the last procedure? Did all groups follow the procedure?

How does this information help you to solve the challenge?"

2. Students' responses should be based on the evidence from the procedures they ran. The class procedure is now more precise, and their data should be more clustered, but some variation probably remains. From this, they can conclude that it is unlikely that they could ever get an exact answer with the materials and conditions in the classroom.
3. Students might say that a more exact measurement could be obtained by using more precise tools to measure the tilt of the plate or the amount of soap. However, they should recognize that all tools have limited precision, and no measurement is ever exact. Students should realize that there is no exact value, but they can still have trustworthy results if, when following the same procedures under the same conditions, their results are repeatable.

TEACHER TALK

"You revised your procedure. How did the new procedures affect the class's results? How consistent are the results? Can you account for any data points that seem too high or too low? What do you think you could do to revise the procedures again and make the data even more consistent?"

Discuss ways in which the procedure can be improved.

TEACHER TALK

"Were there any things groups were doing differently? Could your procedures have been more precise? What else could you do to get more consistent results?"

More to Learn

Lava

viscous: having a sticky, thick, or gluey consistency; not free flowing.

Lava is melted rock that reaches the surface of Earth. There is also melted rock inside Earth. Melted rock is called magma as long as it stays underground and does not reach the surface.

Scientists describe lava by its physical appearance—the way it looks. Several things affect the way lava looks:

- what it is made of (the kinds of molecules and atoms),
- the temperature at which it is flowing,
- the kinds of crystals that form, and
- the land it is flowing over.

When lava comes out of a volcanic vent, it can range in temperature from about 700°C to about 1200°C (1300°F to 2200°F). As it moves away from the vent, it starts to cool. Just like hot fudge on ice cream, lava gets more **viscous**, or thicker, as it cools. Hotter lava is generally thinner than cooler lava. This also means that lava slows down the farther away it gets from the vent where it first erupted.

The composition of lava, or what it is made of, depends on the location of the volcano. You will study more about this later on.

Basaltic Lava

Most lava is basaltic. This type of lava is thin and runny. It has lower silica content than other types of lava. Silica is the same thing sand is made of. The two types of lava commonly found in Hawaii are both basaltic.

A'a

A'a (pronounced "ah-ah") is a slightly viscous lava. This type of lava usually comes out in flows that are three to five meters (approximately 10-16 ft) thick. A'a flows are jagged and extremely sharp. A'a in Hawaiian means "hard on the feet." Even though a'a is viscous, it flows quickly.

An a'a flow appears to eat up cars and a road. The cars give you an idea of how thick the flow is.

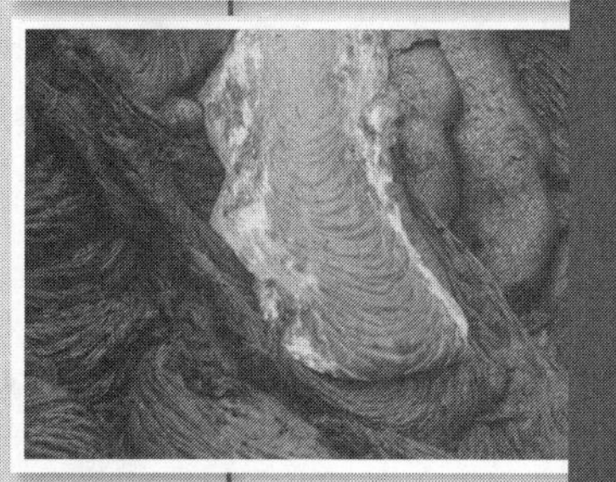

A'a lava as it cools. Even after the lava cools, you would not want to walk on it with bare feet.

△ Guide

Guide students' reading through the types of lava. Let students know they will learn a little about the different types of lava, and where those different types of lava are found. Students could think about how they might revise their procedure to model the worst-case scenario for the challenge of the Hawaiian town near the volcano.

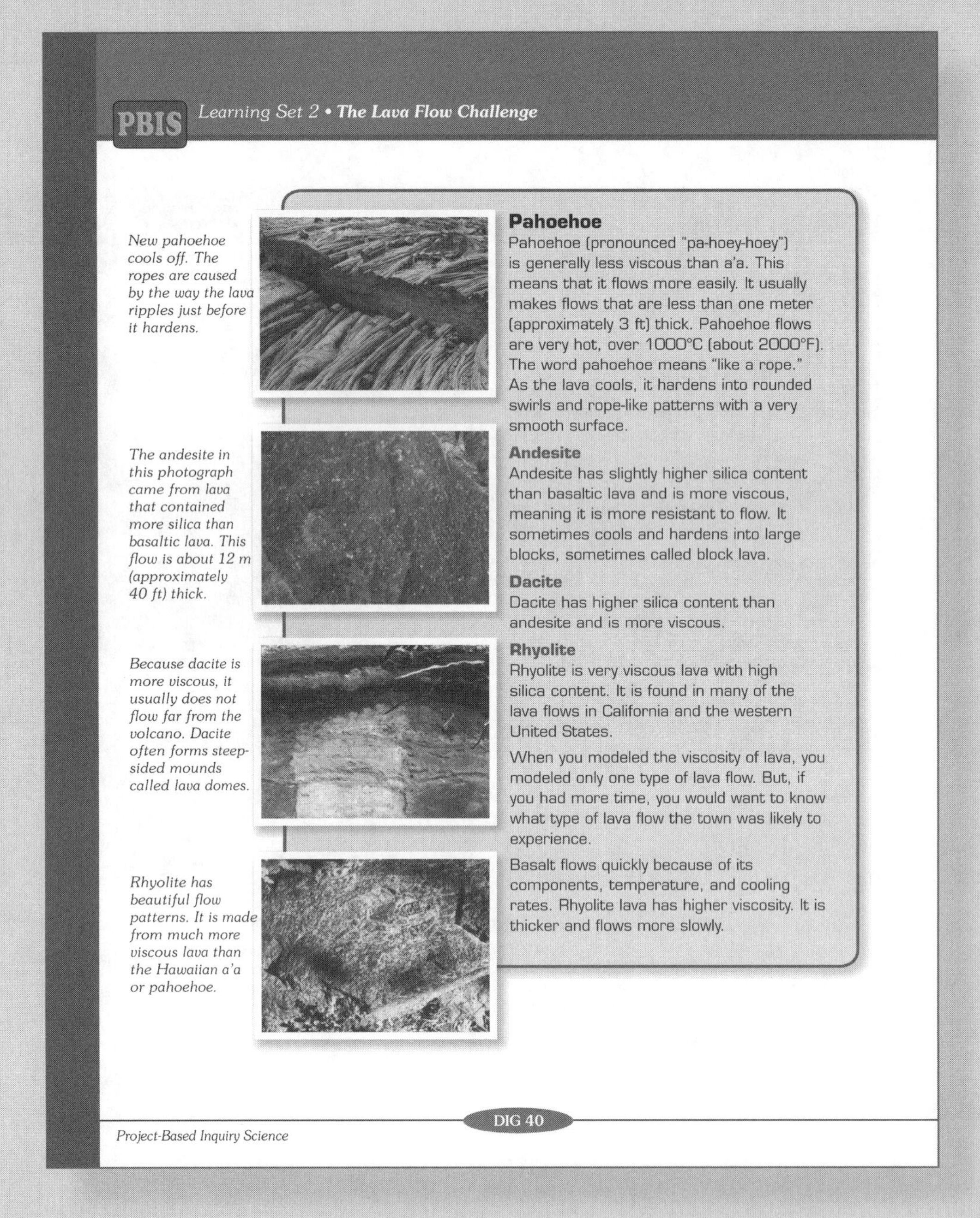

New pahoehoe cools off. The ropes are caused by the way the lava ripples just before it hardens.

The andesite in this photograph came from lava that contained more silica than basaltic lava. This flow is about 12 m (approximately 40 ft) thick.

Because dacite is more viscous, it usually does not flow far from the volcano. Dacite often forms steep-sided mounds called lava domes.

Rhyolite has beautiful flow patterns. It is made from much more viscous lava than the Hawaiian a'a or pahoehoe.

Pahoehoe

Pahoehoe (pronounced "pa-hoey-hoey") is generally less viscous than a'a. This means that it flows more easily. It usually makes flows that are less than one meter (approximately 3 ft) thick. Pahoehoe flows are very hot, over 1000°C (about 2000°F). The word pahoehoe means "like a rope." As the lava cools, it hardens into rounded swirls and rope-like patterns with a very smooth surface.

Andesite

Andesite has slightly higher silica content than basaltic lava and is more viscous, meaning it is more resistant to flow. It sometimes cools and hardens into large blocks, sometimes called block lava.

Dacite

Dacite has higher silica content than andesite and is more viscous.

Rhyolite

Rhyolite is very viscous lava with high silica content. It is found in many of the lava flows in California and the western United States.

When you modeled the viscosity of lava, you modeled only one type of lava flow. But, if you had more time, you would want to know what type of lava flow the town was likely to experience.

Basalt flows quickly because of its components, temperature, and cooling rates. Rhyolite lava has higher viscosity. It is thicker and flows more slowly.

TEACHER TALK

"The reading includes information about lava. Each of the different types of lava flow differently, and that's something to pay attention to as you read this. As you read more about the lava types you should think about which type of lava flows fastest."

More to Learn

Reflect

- How does what you just learned about different types of lava affect the way you might measure lava flow?
- Now that you know more about lava, what would you change in your measurement procedure if you had another chance to revise it?

DIG 41

DIGGING IN

Review the difference between magma and lava. Magma is melted rock inside Earth; lava is melted rock that reaches Earth's surface. Lava moves away from the volcanic vent. As it moves, it cools and moves more slowly. How fast lava flows depends on the type of lava. The more silica in the lava, the more viscous the lava will be, and the more slowly it will flow.

Briefly discuss the types of lava. Emphasize that the two common types of lava commonly found in Hawaii are a'a and pahoehoe.

A'a (pronounced "ah-ah") lava is slightly viscous—not as thin and runny as other types of basaltic lava. A'a flows are typically three to five meters thick and jagged and sharp.

Pahoehoe is less viscous than a'a, but a'a generally flows more quickly. Pahoehoe flows are usually less than one meter thick and cool into rounded swirls and rope-like patterns with smooth surfaces.

Emphasize that a'a lava flows have more volume than pahoehoe. This should cause students to consider increasing the amount of dish soap they are using in their experiment.

Andesite has slightly more silica content and is more viscous than basaltic lava, so it flows more slowly. It sometimes cools and hardens in large blocks. Dacite has more silica content and is more viscous than andesite. Rhyolite has a lot of silica content and is very viscous.

NOTES

Assessment Options

Targeted Concepts, Skills, and Nature of Science	How do I know if students got it?
Scientists often work together and then share their findings. Sharing findings makes new information available and helps scientists refine their ideas and build on others' ideas. When another person's or group's idea is used, credit needs to be given.	**ASK:** How would your conclusions be different if you ran all of the trials by yourself, without your classmates? **LISTEN:** Students may suspect that their trials would have been consistent, but they should recognize that they would not have been able to verify that other researchers would get similar results.
Criteria and constraints are important in design.	**ASK:** How do you know if measurements are reliable? **LISTEN:** Students should explain that if they can be repeated by using the same procedures of measurement, then their measurements are reliable.

Teacher Reflection Questions

- How did students make progress in graphing and analyzing data from the beginning of this *Learning Set* to this section? What evidence do you have that students understand the need for repeatable results?
- How did you encourage students to stay focused on the procedural steps and continue to collect data that was reliable?
- The emphasis in *PBIS* is collaboration. Having students talk to each other is critical. What observations did you make during the class to demonstrate that students were engaged and sharing their ideas with their peers in small group discussions and class discussions? What ideas do you have to encourage student interactions during discussions?

BACK TO THE BIG QUESTION INTRODUCTION

Back to the Big Question

How do scientists work together to solve problems?

Overview

Students reflect on how using standardized procedures helped them to get more repeatable, accurate results. They consider what the sources of the inconsistent data were from the first time they ran their procedures, and identify how they addressed those sources of inconsistent data. They consider how repeatable results are more trustworthy than non-repeatable results. Finally, they build on what they have learned about magma and lava by learning about rocks and minerals in a *More to Learn* segment.

Targeted Concepts, Skills, and Nature of Science	Performance Expectations
Scientific investigations and measurements are considered reliable if the results are repeatable by other scientists using the same procedures.	Students demonstrate understanding that their results were more trustworthy when everyone got similar results. They understand that when everyone gets similar results, the results are repeatable.

Homework Options

Reflection

- **Science Process:** Why is it important for everyone to get similar results? *(Students should state that when everyone gets similar results, they know the results can be repeated. This indicates that the results are trustworthy.)*

BACK TO THE BIG QUESTION IMPLEMENTATION

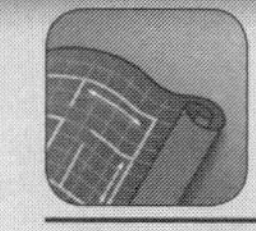

Learning Set 2

Back to the Big Question

How do scientists work together to solve problems?

You and your classmates have been trying to develop an accurate procedure. In the end, you probably realized that it would be very difficult to develop a fully accurate procedure. But, as the different groups in the class used more similar procedures, their answers got closer to one another. You found that the way you collect data affects the answers you can find.

The first time everyone tried to determine how fast the dish soap ran across the plate, each group had different results. That is because each group used a similar, but not identical, method. The class then came up with a more standard procedure. When everyone followed this procedure, the results were closer to each other. Your data became more consistent.

There are three likely sources of inconsistent data:

- Different procedures are used for different trials.
- Factors that can affect the measured result are not carefully controlled.
- The tools used have constraints.

It is important for scientists that the results of their investigations can be trusted. They must develop very precise methods that give similar results each time. To check scientific results, other scientists repeat procedures to see if they get the same results. Scientists can trust the work of other scientists if another scientist can replicate the investigation and get the same results.

Learning Set 2

Back to the Big Question

10 min.

Students reflect on what they have learned and how it connects to the Big Challenge: How do scientists work together to solve problems?

△ Guide

Use the student text to help students connect their experiences in this *Learning Set* to the *Big Question* of the Unit, highlighting the three likely sources of inconsistent data. Pose questions to connect their experiences with the *Lava Flow Challenge* to the *Big Question.*

TEACHER TALK

"What about these investigations helped us to get more reliable results? How did the way you worked together change your investigations? What in these investigations might be like what scientists do?"

META NOTES

Students should say that, like scientists, they designed standardized procedures and worked towards repeatable results. During the investigations, they should have made their procedures more specific and ensured that everyone was using the same procedures. And scientists consult with each other, share their results, and revise their investigations.

More to Learn

Rocks and Minerals

Lava is melted rock, and rock is the material that makes up Earth's solid part. Rocks are composed of **minerals**. Minerals are **inorganic** solids formed in nature, with definite chemical compositions, and unique **crystalline** arrangements of atoms. Inorganic materials do not come from living things but are formed from matter on Earth.

Minerals are either chemical elements or compounds and have specific chemical and physical properties. The properties of a mineral depend on its composition and how it was formed. Several easily observed properties commonly used to identify minerals are color, luster (how it reflects light), streak (color of the powdered mineral), hardness (how easily it is scratched), and cleavage (how it breaks apart).

A rock is a naturally formed, inorganic Earth material that holds together in a firm, solid mass. Most rocks are made of lots of grains that hold together. The grains may be mineral crystals, rock fragments, or even the solid parts of once-living things. Rocks are identified by their mineral composition and **texture**. Texture is the size, shape, and arrangement of the mineral crystals or grains in the rock.

Based on how they are formed, rocks are divided into three main categories: **igneous, sedimentary,** and **metamorphic**.

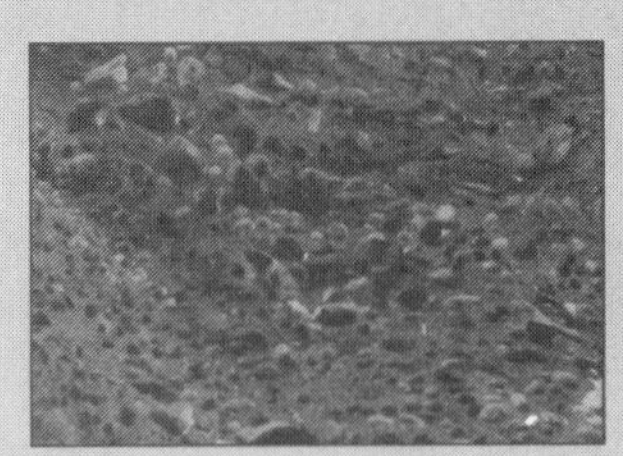

Basalts, igneous rocks formed by the rapid cooling and hardening of lava flows, are the most widespread of the igneous rocks.

Igneous rocks (from the Latin word for fire) form when hot, molten magma or lava cools and becomes solid. As the molten rock cools, mineral crystals form and increase in size until all of the material has

mineral: a naturally formed, inorganic solid, with a definite chemical composition, and a specific arrangement of atoms in a crystalline pattern.

inorganic: material that does not come from living things.

crystalline: describes the arrangement of atoms in a specific structural pattern.

texture: in this case, the size, shape, and arrangement of the grains in rocks.

Engage

Ask students if they have seen rocks formed from lava or magma. They probably have, even if they did not recognize it. Tell students that rocks formed from lava are known as igneous rock, one of the three basic types of rock.

PBIS Learning Set 2 • **The Lava Flow Challenge**

igneous: one of the three main categories of rock, formed when hot, molten magma or lava cools and becomes solid.

sedimentary: one of the three main categories of rock, formed when eroded and deposited rock particles are compacted or cemented together, or from particles left behind as water.

metamorphic: one of the three main categories of rock, formed when already existing rocks are changed by heat, pressure, or chemical action.

pressure: the amount of compression force acting on a substance.

become solid. Usually, the faster an igneous rock cools and solidifies, the smaller its crystals, because they have had less time to increase in size.

Sandstone, a sedimentary rock, is mainly composed of sand-size mineral or rock grains. It forms from sediment accumulating, then compacted and cemented together.

Most sedimentary rocks form when sediments (eroded and deposited rock particles) are compacted or cemented together. Other sedimentary rocks form from particles left behind as water evaporates. Some even form from the shells of marine life. Sedimentary rocks make up less than 10 percent of Earth's crust, but they account for about three-quarters of all surface rock.

Marble, a metamorphic rock, is formed from stone that has been changed by both extremely high temperatures and tremendous pressure.

Metamorphic rocks (from the Greek words *meta*, meaning change, and *morph*, meaning form).These rocks can form from igneous, sedimentary, or even other metamorphic rocks. Great heat and **pressure** can cause the grains in the rock to change in size, shape, and density, and can cause chemical changes in the rock's minerals.

DIG 44

Project-Based Inquiry Science

△ Guide

Tell students that rocks are mostly made of minerals, which are inorganic solids with crystalline structures. Minerals may be elements or chemical compounds. Minerals have specific chemical and physical properties. They are often identified by their color, luster, streak, hardness, or cleavage.

Rocks are made of grains that hold together. The grains may be mineral crystals, rock fragments, or remains of once-living things. Rocks are differentiated by their mineral composition and texture.

TEACHER TALK

"What is a rock? A rock is actually composed of grains. The grains might be mineral crystals, or they might be fragments of older rocks. They might even be pieces of shell, bone, or other bits of living things."

Igneous rocks form from lava (or magma). As lava cools, mineral crystals form. Sedimentary rocks form from particles deposited by water. Sedimentary rocks make up about three-quarters of the rocks on Earth's surface. Metamorphic rocks can form from igneous or sedimentary rocks, they are rocks in which the grains have been changed by great heat and pressure.

TEACHER TALK

"Now we know a little about igneous rocks. Igneous rocks can form from lava, or from magma. As lava or magma cools, crystals form, which together form rocks.

Sedimentary rocks are formed in a completely different way. Sedimentary rocks form when water deposits particles on the ground. This can happen when water from a stream evaporates, for instance.

Either type of rock can be changed by heat and pressure. Rocks that have been changed by heat and pressure are called metamorphic rock."

NOTES

LEARNING SET INTRODUCTION

Learning Set 3

The Basketball-Court Challenge

◀ ***17 class periods***

A class period is considered to be one 40 to 50 minute class.

Student groups investigate the causes of erosion and how to control it. Then they make recommendations to the school board on how to design a basketball court at the bottom of a hill.

Overview

Students investigate what causes erosion and how to control it while being introduced to the tools of *PBIS* and engaging in the social practices of scientists. Students are first introduced to the challenge: to make a recommendation to a school board on how to manage the erosion uphill from a basketball court at the foot of the hill. Students begin by learning about what causes erosion by identifying erosion around their school and working in groups to investigate case studies. Using a model, groups study how the flow of water over sloped land affects the erosion of various types of dirt. Students learn about scientific explanations as they construct explanations of what causes erosion based on their observations and information from the case studies. After learning about the cause of erosion, they begin to investigate how to control erosion. Groups analyze case studies of erosion control and then investigate erosion control with a model. They plan, build, and test their ideas for erosion control for the basketball court using a model. Finally, students make a recommendation based on their findings to the school board.

LOOKING AHEAD

***Sections 3.3, 3.7,* and *3.9* involve using large amounts of Earth materials. You may want to begin considering how you will deal with distribution, collection, and clean up for these materials.**

Targeted Concepts, Skills, and Nature of Science	Section
Scientists often work together and then share their findings. Sharing findings makes new information available and helps scientists refine their ideas and build on others' ideas. When another person's or group's idea is used, credit needs to be given.	3.1, 3.2, 3.3, 3.4, 3.5, 3.6, 3.7, 3.8, 3.9, 3.10
Criteria and constraints are important in design.	3.1, 3.6
Scientists must keep clear, accurate, and descriptive records of what they do so they can share their work with others and consider what they did, why they did it, and what they want to do next.	3.1, 3.2, 3.3, 3.6, 3.7,3.8, 3.9, 3.10

Targeted Concepts, Skills, and Nature of Science	Section
Identifying factors that could affect the results of an investigation is an important part of planning scientific research.	3.2, 3.3, 3.5
Scientific investigations and measurements are considered reliable if the results are repeatable by other scientists using the same procedures.	3.3, 3.6, 3.7
In a fair test, only the manipulated (independent) variable, and the responding (dependent) variable change. All other variables are held constant.	3.3, 3.6, 3.7
Erosion is the process of soil and other particles being displaced by water, waves, wind, and gravity.	3.1, 3.2, 3.3, 3.4, 3.5, 3.6, 3.7, 3.8, 3.9, 3.10
Scientists make claims (conclusions) based on evidence (trends in data) from reliable investigations.	3.2, 3.3, 3.4, 3.7, 3.10
Explanations are claims supported by evidence, accepted ideas, and facts.	3.4, 3.8, 3.9, 3.10
Scientists use models to simulate processes that happen too fast, too slow, on a scale that cannot be observed directly (either too small or too large), or that are too dangerous.	3.3, 3.6, 3.7, 3.9, 3.10

Students' Initial Conceptions and Capabilities

- Students of all ages may hold the view that the world was always as it is now. (Freyberg, P. 1985.)
- Students often think of models as physical copies of reality rather than conceptual representations. (Grosslight et al., 1991.)

Understanding for Teachers

The goal of this Unit is to familiarize students with the nature of science, the *PBIS* curriculum, and some of the tools used in this curriculum. It is not intended for the students to gain a deep understanding of the scientific concepts of erosion.

Erosion

Erosion happens when soil is moved from one place to another by means of wind, water, ice, or animals and the interaction due to gravity. Erosion is different from weathering, which deals with the decomposition of rocks (or Earth materials) with no movement involved. Erosion always deals with the movement of Earth materials. It is a natural process and is often a useful process. Erosion is responsible for creating our beaches, as well as removing them. Erosion is responsible for creating sediments in rivers and moving these down stream to create nutritious soil for vegetation, and sustaining ecosystems. However, it can cause problems. In many places erosion is increased significantly because of the way the land is used.

The rate of erosion depends on precipitation, the texture of the soil, the shape of the land, the ground cover, etc. For more information about erosion types and how to control erosion try an Internet search on the keywords erosion or erosion control.

Models

Models are used when an investigator wants to study a phenomenon of something that is too big, too fast, too slow, too small, or too dangerous to study directly. In this *Learning Set*, modeling is used in three different ways.

In *Section 3.3*, a model of land is used to investigate how erosion by water affects various Earth materials by going through the material or over a saturated hill of the material. The model is used to run an experiment to ascertain how certain factors affect a phenomenon.

In *Sections 3.6* and *3.7*, a model is used and varied to explore how to control the phenomenon of erosion.

A model is used to test students' ideas on how to control the erosion uphill for the *Basketball-Court Challenge*.

NOTES

$1\frac{1}{2}$ to 2 class periods▶*

Learning Set 3

The Basketball-Court Challenge

10 min.

Students are introduced to the challenge of creating a recommendation for controlling erosion near a proposed basketball court.

LEARNING SET 3 IMPLEMENTATION

Learning Set 3

The Basketball-Court Challenge

Imagine that your school sits on top of a hill. There is no room for a basketball court. One of the parents is willing to donate some land at the bottom of the hill for the school to use as a basketball court. Everyone is excited about the idea until the school board meeting.

"Wait a minute," says the school board president. "If we build a basketball court at the bottom of the hill, how are we going to control **erosion**? What is the point of having a basketball court if it is going to be covered in dirt and mud all the time? We do not have the money to spend on a landscape report."

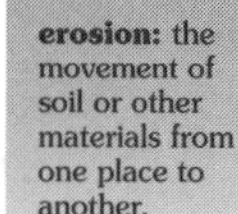

erosion: the movement of soil or other materials from one place to another.

ingenuity: cleverness and originality.

The school board votes not to accept the donation because of the potential erosion problem. You and your friends are very disappointed. You talk to your science teacher, and she says that erosion is a problem that can be controlled. It might take **ingenuity**, but you should be able to figure out something.

Your entire class goes to the next school board meeting. Your teacher proposes that the class investigate erosion and suggest ways it can be controlled. The school board agrees to this plan. The next school board meeting is in three weeks. If you can provide them with enough data to make an informed decision about managing erosion, they will consider accepting the land donation to build the basketball court.

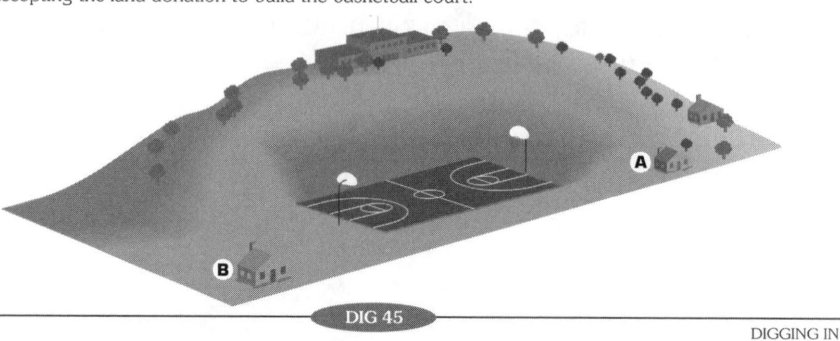

DIG 45 DIGGING IN

○ Engage

Motivate students and introduce the challenge by discussing why it would be desirable to have a basketball court on school grounds.

TEACHER TALK

"It is amazing how popular basketball is. How many of you like shooting hoops at lunch or after school? Imagine you were at a school that didn't have a basketball court. Would you miss the chance to play?"

*A class period is considered to be one 40 to 50 minute class.

Union School District
Mary Chalmers, Superintendent
27 Courthouse Square
Springfield, IL 62700

Dear Science Class:

We are delighted that your class has taken on the challenge of studying erosion control around the proposed basketball court. We will expect your report within two weeks with recommendations for erosion control on the hill above the court.

The land in question is an empty lot at the foot of a hill. There are houses on both sides of the site. You will have to make sure that none of your erosion-control measures affect these houses. We look forward to you presenting your recommendations to the school board at our next meeting.

Please use the information we have included to help you understand the requirements. Remember, you will have to be careful not to cause any damage to the properties on either side of the court.

The Proposed Basketball Court

Dimensions of court	length 28 m (about 92 ft)
	width 15 m (about 49 ft)
Vertical height of hill	10 m (about 33 ft)
Required distance from base of hill to court	5 m (about 16 ft)
Distance to houses on either side	12 m (about 39 ft)
Size of houses	30 x 10 m (about 98 x 33 ft)

The school board is grateful that you are able to take on this project. Good luck, and let us know if you need more information.

Sincerely,

Mary Chalmers

Mary Chalmers
School Superintendent

TEACHER TALK

“The school described in this challenge does not have a basketball court but the students would like one. There is some land near the school that is a good size for a court. The school board is thinking of using the land for a basketball court.”

TEACHER TALK

"The problem is that the land slopes down and the basketball court would sit at the bottom of the slope. That's where all the rain that falls on the hill goes. The water carries a lot of dirt as it flows over the ground so the bottom of the hill is wet and sometimes covered with dirt. The only way to convince the school board to build the basketball court would be to prevent it from flooding and prevent all the dirt moving down onto it."

Read the scenario and the letter in the student text. Make sure students understand the challenge is to find a way to prevent erosion from ruining the proposed basketball court without damaging any neighboring properties. If they cannot meet the challenge, the basketball court will not be built. Consider drawing diagrams to illustrate the scenario if you think it will help students understand.

Point out that this challenge is very much like real-world challenges, for which engineers devise a wide variety of solutions. Like engineers, students will need to use their ingenuity to solve the problem and to devise unique solutions.

NOTES

SECTION 3.1 INTRODUCTION

3.1 Understand the Challenge

Thinking About Erosion

◀ ***$1\frac{1}{2}$ to 2 class periods***

A class period is considered to be one 40 to 50 minute class.

Overview

Students are introduced to the challenge of the *Learning Set*: They will try to find how to prevent erosion deposition at the bottom of a hill near the school so that a basketball court can be built. To solve this problem, they first identify erosion on the grounds of their school and determine what they know about erosion and what they need to learn. They are introduced to the *Project Board*, which is used to organize their ideas and track their progress with the challenge. Finally, they update the class *Project Board*, which prepares them to investigate erosion further.

Targeted Concepts, Skills, and Nature of Science	Performance Expectations
Scientists often work together and then share their findings. Sharing findings makes new information available and helps scientists refine their ideas and build on others' ideas. When another person's or group's idea is used, credit needs to be given.	Students should work with their groups to decide which examples of erosion they can agree on and which they want to learn more about.
Criteria and constraints are important in design.	Students should identify criteria and constraints with the class.
Scientists must keep clear, accurate, and descriptive records of what they do so they can share their work with others and consider what they did, why they did it, and what they want to do next.	Students should keep track of what they are learning and what they need to learn using the *Project Board*. They should also record their observations using sketches and notes.
Erosion is the process of soil and other particles being displaced by water, waves, wind, and gravity.	Students should observe erosion around the school and make conjectures about what caused the erosion.

Materials	
1 per student	*Erosion-Walk Observation* pages
1 per student (Optional)	Digital camera Photocopy of erosion images in the student text, for students to refer to during the Erosion Walk
1 per class	Class *Project Board*

Activity Setup and Preparation

Determine where you will take students for their erosion walk prior to class. Look to walk around the school grounds for signs of erosion.

This section lasts two class periods. A good place to have a break between class periods is just after the erosion walk or the *Conference* segment, and before the *Create the Project Board* segment.

Homework Options

Reflection

- **Science Content:** Based on the examples of erosion you observed, in what kind of environment would you expect to see the worst erosion? Describe it. *(Students should describe an environment that combines the factors that lead to heavy erosion: an environment with sloping land and heavy rainfall. There are many other factors that could be listed: sparse vegetation, fine soil, and winds among them.)*
- **Science Process:** How did looking at erosion around school help you better understand what causes erosion and determine what you need to learn? *(This question is intended to get students thinking about effective ways of solving problems and learning.)*

Preparation for 3.2

- **Science Content:** What are some of the things your group decided they need to learn about erosion? How do you think you could find the answer? *(Students' answers should reflect some of the topics that were discussed in class.)*

SECTION 3.1 IMPLEMENTATION

3.1 Understand the Challenge

Thinking About Erosion

You are going to begin by identifying the criteria and constraints of this challenge. You will then walk around your school, looking for examples of erosion. Understanding what erosion is and what causes it will be important for addressing the challenge. For now, think about erosion as movement of soil or other ground material from one place to another.

Identify Criteria and Constraints

It is always a good idea, before beginning to address a challenge, to make sure you understand the challenge. One way to do that is to identify the criteria and constraints. Remember that criteria are what you need to accomplish, and constraints are limitations on your solution. For the *Basketball-Court Challenge*, your criteria are what you need your erosion-control method to be able to do. The constraints on your solution are what you have to keep in mind and be careful about as you work on a solution. Review the letter that you received from the school superintendent, and record the criteria and constraints you identify. Then, as a class, list and discuss the criteria and constraints for this challenge.

Once you are aware of the criteria and constraints for your design challenge, you can decide which ideas are worth spending more time on and which ones are not.

The Basketball-Court Challenge	
Criteria	Contraints
The solution will keep mud and dirt from sliding down the hill onto the basketball court	

3.1 Understand the Challenge

Thinking About Erosion

5 min.

Students are introduced to erosion and what it means for the challenge.

△ Guide

After reading the letter in the *Learning Set* introduction, tell students that to understand the challenge they will need to know what erosion is. Emphasize that erosion is the process in which soil and other particles are moved from one place to another by wind, water, or ice, and gravity.

Identify Criteria and Constraints

10 min.

Students identify the criteria and constraints for the Basketball-Court Challenge.

...her ground ma...

Identify Criteria and Constraints

It is always a good idea, before beginning to address a challenge, to make sure you understand the challenge. One way to do that is to identify the criteria and constraints. Remember that criteria are what you need to accomplish, and constraints are limitations on your solution. For the *Basketball-Court Challenge*, your criteria are what you need your erosion-control method to be able to do. The constraints on your solution are what you have to keep in mind and be careful about as you work on a solution. Review the letter that you received from the school superintendent, and record the criteria and constraints you identify. Then, as a class, list and discuss the criteria and constraints for this challenge.

Once you are aware of the criteria and constraints for your design challenge, you can decide which ideas are worth spending more time on and which ones are not.

The Basketball-Court Challenge

Criteria	Contraints
The solution will keep mud and dirt from sliding down the hill onto the basketball court	

DIG 47 DIGGING IN

META NOTES

Students may provide a definition of erosion by example, such as, "Erosion happens when water carries dirt downhill." Students may not be able to clearly state more than this. The definition of erosion will be revised informally throughout the *Learning Set*.

Read the text if students have continued difficulty defining erosion. Students will be finding examples of erosion on the erosion walk so they will be able to further define erosion based on those results. Emphasize that erosion is the process in which soil and other particles are moved from one place to another by wind, water, or ice, and gravity.

△ Guide

Initiate a discussion of the importance of understanding the criteria and constraints of the challenge. Review the meaning of criteria and constraints. Criteria are the conditions that must be met and constraints are limitations in how the criteria can be met.

TEACHER TALK

"We have discussed the problem—if you put a basketball court at the bottom of the hill, water will run down the hill and eroded dirt, carried by the water, will cover the basketball court. We want to find a way to build the basketball court without that happening. We need to identify the goals and limits of the challenge. What areour criteria and constraints? Remember criteria are the conditions that must be met and constraints are the limitations."

Review the letter sent to the students, if necessary. Use the information in the letter to remind students of the criteria and constraints of the challenge. As students identify criteria and constraints, create a class record so that students can refer to the list as they work on their solutions.

◇ Evaluate

Before proceeding, make sure these criteria and constraints are in the class list:

Criteria: to prevent erosion at the top of the hill from covering a 28 m × 15 m basketball court at the bottom of the hill

Constraints: the basketball court must be 5 m from the base of the hill; the project must not damage houses (30 m × 10 m) that are 12 m from the court on either side; the height of the hill is 10 m.

META NOTES

Understand the Challenge **sections generally include identifying criteria and constraints. These sections help the students better understand the challenge so that they might address the challenge effectively.**

NOTES

The Erosion Walk

25 min.

Students walk around school grounds to identify examples of erosion and to consider their cause.

The Erosion Walk

You will be taking a walk around your school and looking for examples of erosion. Look for things you might not usually notice. These might include a pile of pebbles on the side of the road or small gullies formed by a recent rainfall. You do not have to go very far to find examples of erosion.

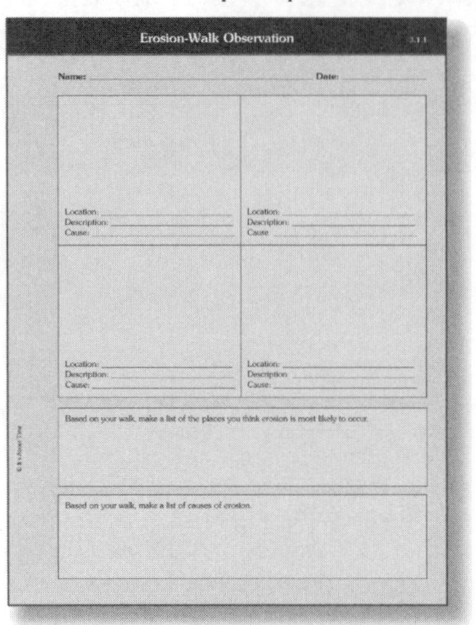
Erosion-Walk Observation

Name: Date:

Location:
Description:
Cause:

Working with a partner, identify at least five examples of erosion. Also identify at least two examples of places where erosion should have happened but did not. Try to find examples that other students have not noticed.

It helps to notice small details. Look for clues, such as dirt that looks different than other dirt around it or rocks piled together. It is important to remember that the effects of erosion can be small or large. Erosion may cover only a few inches of ground, or you might see large areas where soil is washed away. Look carefully as you walk around.

You should also look for places where erosion should have happened but did not. For those places, try to identify what prevented the erosion. If you have any questions about how erosion happens or does not happen, be sure to record them so you will remember.

Recording Your Observations

Record your examples on *Erosion-Walk Observation* pages. You will be able to fit four observations on each page, so each pair of students will have room to record eight examples. For each example, make a sketch of the eroded area. If you can see where the eroded material came from, your sketch should also include the path the material traveled and what was formed when it was deposited. Record the location of your example, and describe it in words. Try to figure out what caused the erosion, and record that as well. You may need to look closely at the area around where you found the example of erosion to figure out how it happened. For your examples of where erosion did not happen, figure out and record what

○ Engage

Probe students to describe where they have seen examples of erosion. Ask them to consider why the erosion may have happened.

△ Guide

Introduce the erosion walk. Suggest that erosion is not just a problem near rivers or oceans, but can be found everywhere. They will now walk around the and look for examples of erosion. Ask students to think about what they might see near the school that would indicate erosion. Students should suggest piles of pebbles, small gulleys, and any places where the soil has a different color from the surrounding soil.

Students will be working in pairs to find five examples of erosion (if the erosion walk path is paved, have students find three examples). Students should make a sketch of each example of erosion they see, record where they saw it, and indicate on their sketches where the eroded material came from and where it was deposited. For each of the expected examples of erosion that they did not see, they should explain why the erosion did not occur. Each partner will have an *Erosion-Walk Observation* page, so they can record eight observations.

⬡ Get Going

Distribute the *Erosion-Walk Observation* pages, and lead the students through a walk. Bring digital cameras if you have them and use them to record erosion evidence.

△ Guide and Assess

As students identify examples of erosion and make their sketches, encourage them to pay attention to small details. Look at students' sketches and make sure they are capturing these details. Make sure they record ideas about why the erosion occured.

META NOTES

When students experience science first-hand, they are able to connect the science knowledge more readily to their lives. It would be possible to send students home with this assignment for homework, but it would then lose the collegial and sharing aspects.

By doing the Erosion Walk as a class, students will have similar experiences and will be able to discuss, throughout the Unit, this common event.

NOTES

Conference

10 min.

Groups discuss their observations and identify what they think they know and what they need to learn about how erosion works.

META NOTES

During a conference, members of a group discuss their ideas. After *the Erosion Walk*, students will share different occurances of erosion they found. Articulating their ideas helps students refine their thinking. Later, during class discussions, groups' ideas are further refined. In this case they begin thinking about what causes erosion.

META NOTES

Some very interesting examples of erosion might be too difficult for students to explain. Encourage students to share the examples that interest them whether they have an explanation or not.

prevented the erosion. Record enough so that you will be able to share your observations with others. You and your partner may agree or disagree. If you disagree, write enough so that you remember what you disagreed about.

You will need a hard surface to write on as you complete your *Erosion-Walk Observation* page, so take a book or workbook with you to lean on. If you have a camera, bring that too.You should also have a pencil and your *Erosion-Walk Observation* page.

Conference

After you return from the erosion walk, share your observations and ideas with your group about what caused your examples of erosion. Make sure everybody has a chance to share. As a group, select two examples of erosion, and make your best guess about how each happened. For example, you may have found a small ditch carved out in a flat area. You might think that the wind slowly carried particles of dirt away from the area. Your guess may not be correct, but do your best based on what you know so far.

Then select an example of a place where erosion did not happen. Try to come to an agreement about why it did not occur.

When you are finished, try to answer the two questions at the bottom of your *Erosion-Walk Observation* page.

- Where is erosion most likely to happen?
- What causes erosion?

In the time you have left, discuss which of your answers you are sure about and which you are less sure about. Discuss what you think you still need to learn to fully understand your observations.

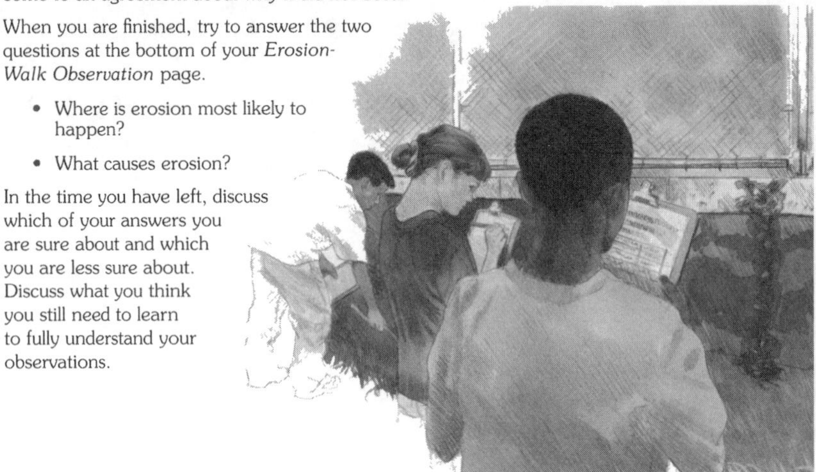

△ Guide

When the class returns from their erosion walk, have each pair meet with another pair to form a group of four. Groups should share all the examples they found and select two examples of erosion to share with the whole class. They should explain why the erosion occurred. They should select an example of a place where they expected erosion, but did not see it and try to explain why it did not occur. Finally, groups should discuss which of their answers they are unsure of and what they need to learn to be sure of their understanding.

When students are finished discussing why erosion did not occur where they expected, they should fill out the two spaces at the bottom of their *Erosion-Walk Observation* pages. The answers to these questions will be addressed during the *Project Board* discussion that follows.

META NOTES

Section 3.1 **requires at least two class periods. This is a good place to end a class if you will not be able to complete the** ***Project Board*** **discussion in the class.**

☐ Assess

As groups discuss the examples they collected, monitor how they share ideas and what they disagree about. Note disagreements—these should be discussed as things to investigate when you create the *Project Board.*

NOTES

Create the Project Board

35 min.

Students share ideas about erosion so they will be able to address the Basketball-Court Challenge and learn to use a Project Board *for monitoring progress on their challenge.*

Create the *Project Board*

When you work over a long period of time on a project, it is important to keep track of what you have accomplished and what you still need to do. Throughout *PBIS*, you will be using a *Project Board* to do that. A *Project Board* gives you a place to organize your ideas, what you need to investigate, and what you are learning. Reading the box, *Introducing the Project Board*, will give you a better idea of what a *Project Board* is and what you will use it for.

Project Board: a space for the class to keep track of progress while working on a project.

To get started on this *Project Board*, identify the important science question you need to answer. To design erosion-control measures for a basketball court, you need to understand the answer to these questions: What causes erosion and how can it be controlled? Write these questions on your *Project Board*. Your challenge is to design an erosion-control plan to manage erosion around the proposed basketball court. Add the challenge to the top of your class *Project Board*: *How can erosion around the proposed basketball court be controlled?*

The erosion walk was meant to help you recognize what you understand about erosion. It also helped you think about what you do not understand well enough. These are the things you will record in the first two columns of the *Project Board*.

What causes erosion and how can it be controlled?
How can erosion around the proposed basketball court be controlled?

What do we think we know?	What do we need to investigate?	What are we learning?	What is our evidence?	What does it mean for the challenge or question?

Engage

Connect the erosion walk to the challenge and help students understand that as engineers and scientists work at challenges, they need to keep a record of their ideas so they can monitor their progress. Tell them they will be doing this using a *Project Board.*

TEACHER TALK

"You found a lot of examples of erosion on the erosion walk. Examples of erosion are everywhere. Remember, there are criteria and constraints to this challenge. The goal is to find a way to control erosion so that it does not damage or disturb the proposed basketball court. We also have to keep in mind that we do not want to damage the other properties nearby. We are just getting started and you have a lot of ideas and questions to test before we get close to a solution. We will need a way to keep track of our ideas and progress just as scientists and engineers do as they are addressing challenges. We'll keep track of our ideas using a *Project Board*."

⬡ Get Going

Begin students on the *Project Board* by starting with the question: *What causes erosion, and how can it be controlled?* Emphasize that they need to be able to answer this question for the school board to build the basketball court. Write this question across the top of the *Project Board* or have a student write it for the class across the top of the *Project Board*.

NOTES

META NOTES

Students do not think about science or engineering as dynamic processes that changes as new information becomes available. They may be surprised to hear that they will be returning to the *Project Board* and updating it over time.

It is helpful to keep track of the date items were posted. That way, students can monitor changes in their ideas and their progress on the challenge.

Linking items in different columns that are related to each other with arrows helps students to see how ideas and questions are connected.

During the discussion, listen for students' ideas about what causes erosion and their reasoning.

△ Guide

Move to the first column and ask the students what they think they know that will help them to address the challenge. There is no need to figure out right now if these ideas are right or wrong.

Include some ideas about what they need to investigate. Use the second column for their questions. Includes questions about ideas students do not agree on or have a hard time accepting. Students will disagree about some of the items in the first column. These are indications of questions that need to go in the second column. Things students are surprised by should go in the second column.

- What did you list as things you think you know?
- What did you talk about in your group that everyone agreed on?
- What do you know about erosion from your erosion walk?
- What did you list as things you were unsure about?
- What did you talk about that everyone did not agree on?

What causes erosion and how can it be controlled?
How can erosion around the proposed basketball court be controlled?

What do we think we know?	What do we need to investigate?	What are we learning?	What is our evidence?	What does it mean for the challenge or question?
Erosion is when soil is moved from one place to another by wind, water or ice as a result of gravity. March 30 Erosion is more likely to occur where there is bare soilthan when it is covered with vegetation. March 30 Erosion is more likely to happen where there is slope. March 30 Erosion is more likely to happen where water flows. March 30	Does erosion affect gravel more or less than soil? March 30 What determines where rainwater will flow? March 30 How does water cause erosion? March 30			

◇ Evaluate

With students, look over the ideas and questions on the *Project Board*. Make sure all students have had the opportunity to contribute and their ideas are represented. There should be questions about what causes erosion or how wind, water, or ice cause erosion. Do not expect students to be experts at this point.

NOTES

Be a Scientist: Introducing the *Project Board*

35 min.

Introduce the ProjectBoard *and why it is used.*

META NOTES

The most important feature of the *Project Board* is the discussion that occurs as the students' learning is documented. How students' ideas are accepted, questioned, supported, and clarified will determine how successfully the Project Board is integrated into the classroom culture.

The *Project Board* segments generally require the inclusion of certain questions or content to move the Unit discussion forward.

Be a Scientist

Introducing the *Project Board*

When you work on a project, it is useful to keep track of your progress and what you still need to do. You will use a ***Project Board*** to do that. It gives you a place to keep track of your scientific understanding as you make your way through a Unit. It is designed to help your class organize its questions, investigations, results, and conclusions. The *Project Board* will also help you to decide what you are going to do next. During classroom discussions, you will record the class's ideas on a class *Project Board*. At the same time, you will keep track of what has been discussed on your own *Project Board* page.

The *Project Board* has space for answering five guiding questions:

- What do we think we know?
- What do we need to investigate?
- What are we learning?
- What is our evidence?
- What does it mean for the challenge or question?

Each time you use the *Project Board*, you will record as much as you can in each column. As you work through a Unit, you will return over and over again to the *Project Board*. You will add more information and revise what you have recorded. Everything you write in the columns will be based on what you know or what you have learned. In addition to text, you will sometimes want to put pictures or data on the board.

What Do We Think Know?

In the first column of the *Project Board*, you will record what you think you know. As you just experienced, some things you think you know are not true. Some things are not completely accurate. It is important to record those things anyway for two reasons:

- When you look at the board later, you will be able to see how much you have learned.
- Discussion with the class about what you think you know will help you figure out what you need to investigate.

△ Guide

Explain that the *Project Board* is a tool that will be used throughout the course. It is used by the class to organize ideas, questions, and answers when working on a challenge. The *Project Board* helps keep track of all the things they learn in the Unit.

Emphasize that the class will be filling out the first two columns (*What do we think we know?* and *What do we need to investigate?*) today and they will add to all the columns as they work on the challenge.

Describe the five columns. Explain that the third column lists the claims and the fourth column lists the evidence that backs up those claims based on observations and information from experts. The fifth explains how it is connected with the challenge or a bigger question.

TEACHER TALK

"I have set up our class *Project Board* for the *Basketball-Court Challenge*. All of you have ideas about how erosion works. You probably discovered on the erosion walk that you already know some things. Other things you might not be sure about. Our *Project Board* will help us organize our ideas as we work on this challenge. We will need to answer many questions as we work on the bigger question of the challenge. A *Project Board* is a way to keep track of what we think we know, what we want to investigate, what we are learning and how we know it, and how it all hooks up with the addressing the challenge. (Show the headers on the *Project Board*). We will be returning to our class *Project Board* and updating it many times while we work on our challenge."

NOTES

META NOTES

Students will develop a deeper understanding of what belongs in each column of the *Project Board* as they continue to update it. It is not necessary that students understand how the *Project Board* will be used at this point. There will be many opportunities for them to gain more knowledge of how they can use the *Project Board.*

META NOTES

The first column includes the word think. The intention is to imply the tentative nature of science knowledge, even the science knowledge they come to class with.

What Do We Need to Investigate?

In this column, you will record what you need to learn more about. During your group conference, you probably came up with questions about how to explain some of your observations. You might also have figured out some things you are confused about. And you might have found that you and others in your group disagreed about some things. This second column is designed to help you keep track of things that are confusing. Record what you do not understand well yet and what you disagree about. These are the things you will need to investigate. They will be important for achieving your challenge (designing a method to control erosion).

Sometimes you are unsure about something but do not know how to word it as a question. One of the things your class will do together around the *Project Board* is to turn what you are curious about into questions you can investigate.

Later in this Unit, you will return to the *Project Board*. For now, work as a class and begin filling in the first two columns.

What's the Point?

You observed examples of erosion in your schoolyard or nearby neighborhood. Some may have been small and hard to notice and others may have been large and quite visible. They were caused by different forces in nature, such as running water, wind, and gravity.

You started a *Project Board* to help you keep track of what you understand. You also added some questions and ideas you need to investigate further. The *Project Board* is a space to help the class work together to understand and solve problems. Using it will help you have good science discussions as you work on a project.

Now that you have identified the questions you need to answer, you know what you need to do next. You need to investigate to find the answers to some of those questions.

NOTES

Assessment Options

Targeted Concepts, Skills, and Nature of Science	How do I know if students got it?
Scientists often work together and then share their findings. Sharing findings makes new information available and helps scientists refine their ideas and build on others' ideas. When another person's or group's idea is used, credit needs to be given.	**ASK:** Why was it important to discuss your ideas about erosion with the other members of your group? **LISTEN:** Students should recognize that they needed to discuss their ideas about erosion with their groups to see where they might need to understand erosion better.
Criteria and constraints are important in design.	**ASK:** Why was it important to identify the criteria and constraints of the challenge? **LISTEN:** Identifying the criteria and constraints should help students fully understand the challenge, and may help them begin to see solutions.
Scientists must keep clear, accurate, and descriptive records of what they do so they can share their work with others and consider what they did, why they did it, and what they want to do next.	**ASK:** How did you use your observations? **LISTEN:** Students should have used their observations during their group discussions to determine what they need to understand better.
Erosion is the process of soil and other particles being displaced by water, waves, wind, and gravity.	**ASK:** How would you describe what erosion is? **LISTEN:** Students' descriptions should specify that erosion is when soil and other materials are moved from one place to another by wind, water, or ice.

Teacher Reflection Questions

- What difficulties do you expect students to have understanding erosion? What can you do to help them with these difficulties?
- What evidence do you have that students effectively identified what they do not yet understand about erosion? What can you do to help them identify what they need to learn?
- How were you able to manage the class's time during the erosion walk? What could you try in future erosion walks to make time management easier?

NOTES

SECTION 3.2 INTRODUCTION

3.2 Case Studies

What Causes Erosion?

◀ *$2\frac{1}{2}$ class periods*

A class period is considered to be one 40 to 50 minute class.

Overview

Students begin exploring the causes of erosion by analyzing case studies. The class is introduced to what case studies are. Then the class investigates a case study together to provide a good example of how to analyze and use information from a case. Student groups then investigate one of three cases. Groups then share their analysis via a poster and presentation to the class. At the end of all the presentations, the class discusses the various types of erosion and students reflect on what erosion is, what might affect it, and how it may be controlled. Groups then update the class *Project Board*, filling in what they have learned about erosion and the evidence that supports what they have learned.

Targeted Concepts, Skills, and Nature of Science	Performance Expectations
Scientists often work together and then share their findings. Sharing findings makes new information available and helps scientists refine their ideas and build on others' ideas. When another person's or group's idea is used, credit needs to be given.	Students should work in small groups analyzing case studies and then share their results with the class, students' to on each other's results and ideas.
Scientists must keep clear, accurate, and descriptive records of what they do so they can share their work with others and consider what they did, why they did it, and what they want to do next.	Students should keep records of the case study they are working on and use their records to create a poster and presentation to the class.
Identifying factors that lead to variation is an important part of scientific investigation.	Students should identify different causes for erosion.
Erosion is the process of soil and other particles being displaced by water, waves, wind, and gravity.	Students should be able to describe the causes of erosion after analyzing case studies of erosion.

Targeted Concepts, Skills, and Nature of Science	Performance Expectations
Scientists make claims (conclusions) based on evidence (trends in data) from reliable investigations.	Students describe what causes erosion in their *Project Board* and the evidence from the case studies that support this claim.

Materials	
5 per student	*Erosion Case Study* page
1 per class	Class *Project Board*

Homework Options

Reflection

- **Science Process:** How do case studies help scientists and engineers answer questions? *(Students should describe how analyzing different cases of situations provides scientists and engineers with information that allows them to find trends and draw conclusions.)*
- **Science Content:** What are some of the causes of erosion? How do you think you can stop them? *(Students' responses should include at least two of the following: wind, water, rain. The second part of the question is to get students thinking about how they will try to stop or slow down different causes of erosion.)*

Preparation for 3.3

- **Nature of Science:** How do you think water affects the erosion of soil? Do you think different soils are affected differently by water? *(Students should describe that moving water can erode or move soils and different soils probably are eroded at different rates.)*

SECTION 3.2 IMPLEMENTATION

◂ $2\frac{1}{2}$ *class periods**

3.2 Case Studies

What Causes Erosion?

Looking at evidence of erosion has helped you identify some of the causes of erosion. It has also helped you raise questions about what causes erosion. Other people have had to deal with erosion before. You are going to review some real-life situations in which erosion created big problems. These situations are known as **cases**. The cases you will read about have been studied by scientists and engineers. Those experts have written about these cases so others can learn how erosion happens and how to prevent it. The **case studies** you will be reading are about large-scale erosion. They will help you answer some of your questions about erosion. They will also give you some more ideas about how the erosion you identified might have happened.

case: an example or occurrence of something.

case study: an analysis of an example or occurrence of something.

Be a Scientist

Learning from Cases

Scientists and engineers often face problems that other people have faced before. Sometimes, examining those cases can help suggest ways of dealing with a new problem. A case is an example of something.

When experts work on a case that can teach others something, they often write it up as a case study. A case study includes a description of what happened, the expert's best explanation of why it happened, and lessons others can take from the experience. Cases can often help you see your problem in a new way. They can also show you how others solved similar problems. Cases can sometimes help you understand that your problem cannot be solved in the way you thought it would be. People read about cases to get ideas. Cases provide suggestions to think about and solutions to consider.

3.2 Case Studies

What Causes Erosion?

10 min.

Students are introduced to case studies and why they are useful.

○ Engage

Ask students to describe some causes of erosion they have seen or heard about. The goal is to get some stories of erosion from the students. These will be used as examples of what a case study is.

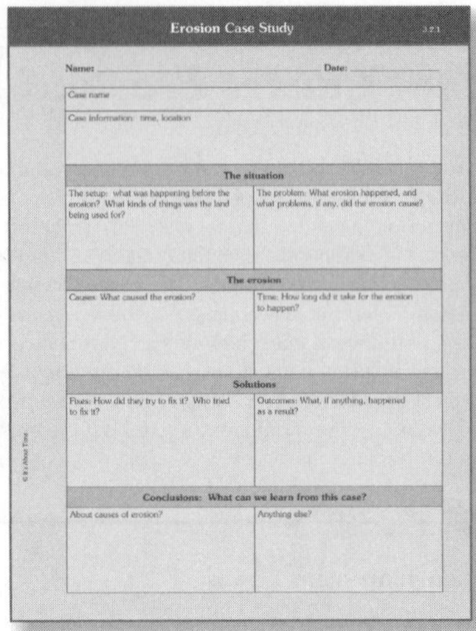

Erosion Case Study 3.2.1

Name: ____________ Date: ____________

Case name	
Case information: time, location	
The situation	
The setup: what was happening before the erosion? What kinds of things was the land being used for?	The problem: What erosion happened, and what problems, if any, did the erosion cause?
The erosion	
Causes: What caused the erosion?	Time: How long did it take for the erosion to happen?
Solutions	
Fixes: How did they try to fix it? Who tried to fix it?	Outcomes: What, if anything, happened as a result?
Conclusions: What can we learn from this case?	
About causes of erosion?	Anything else?

META NOTES

Before students begin reading the case studies, it is important for them to understand what a case study is and how it is used.

Procedure

40 min.

Students learn to analyze a case study by looking at the first case as a class, then reading and analyzing the case studies with their groups.

Procedure

The case studies that follow will give you some ideas about how and when erosion happens. Later, you will read more about how people have tried to reduce or limit the impact of erosion.

You identified some causes of erosion after your erosion walk. Reading these case studies will help you identify more causes of erosion.

1. Read the case study (or case studies) your teacher assigns you.
2. For each case study you are assigned, identify what happened and why. Use the prompts on your *Erosion Case Study* page to help you describe the case you are reading about well enough so others can learn from what you write.
3. Make a list of the causes of erosion that you have read about.
4. Identify an example of erosion you saw on your erosion walk that is similar to the erosion you read about. Use the case you read about to give you ideas about what caused the erosion you observed.

△ Guide

Point out that their descriptions of erosion are basically stories about what they saw. Let them know that scientists and engineers often use these types of stories to try to figure out answers to their questions. Many of these stories are about problems similar to the *Basketball-Court Challenge*. Ask students how looking at these stories can help them learn about erosion.

TEACHER TALK

"On the erosion walk, you recorded erosion that you saw and what you thought happened. You told the story of the erosion that happened there. Scientists and engineers use stories to try to understand things. How do you think looking at stories like this helps scientists? How can stories help you to learn about erosion?"

Emphasize that when you read a story about how something happened, you have some idea of what might happen in similar situations. The more documented stories you read showing similar trends, the more reliable the evidence that these trends will occur for similar situations.

Tell students that scientists and engineers call these cases. A case study is someone's analysis of a case. A case study gives students an idea of what might happen in similar situations, and helps them understand why things happened as they did. Let students know they will look at several case studies, to help solve the *Basketball-Court Challenge*.

META NOTES

In *PBIS*, case studies can be used for the purpose of helping students generate questions or helping them glean ideas or answer questions. Case studies are divided among the class with each group reading and analyzing one case. Usually there are two groups assigned to each case. Students then hold an *Investigation Expo* or share in a way so that they all learn from each case study.

△ Guide

Briefly go through the procedure with the class, discussing each step.

TEACHER TALK

"We are going to go through a case study together. Before we do, let's go over what we will need to do when we analyze a case study. Before reading the case study we should have some idea of what we are looking for. These are case studies of erosion. When you read them you will want to identify what happened in the erosion story or case that you are reading and why it happened. You'll want to make a list of all the causes of the erosion in the case. You should identify an example of erosion from your erosion walk that was similar to the erosion in your case study and describe how it is similar and/or different from the case study you read."

Discuss the *Erosion Case Study* page with the students and how it will help them analyze their case study. Explain why it is important to consider what the land was used for before the erosion occurred and how this could affect erosion. Erosion could have increased after changing how the land was used. In these situations, knowing how the land was may provide insights in to how to control erosion.

Provide an example of how to analyze case studies by going through the first case study with the class. As you read, help students connect the pictures with the text to see where to look for answers to the *Erosion Case Study* questions.

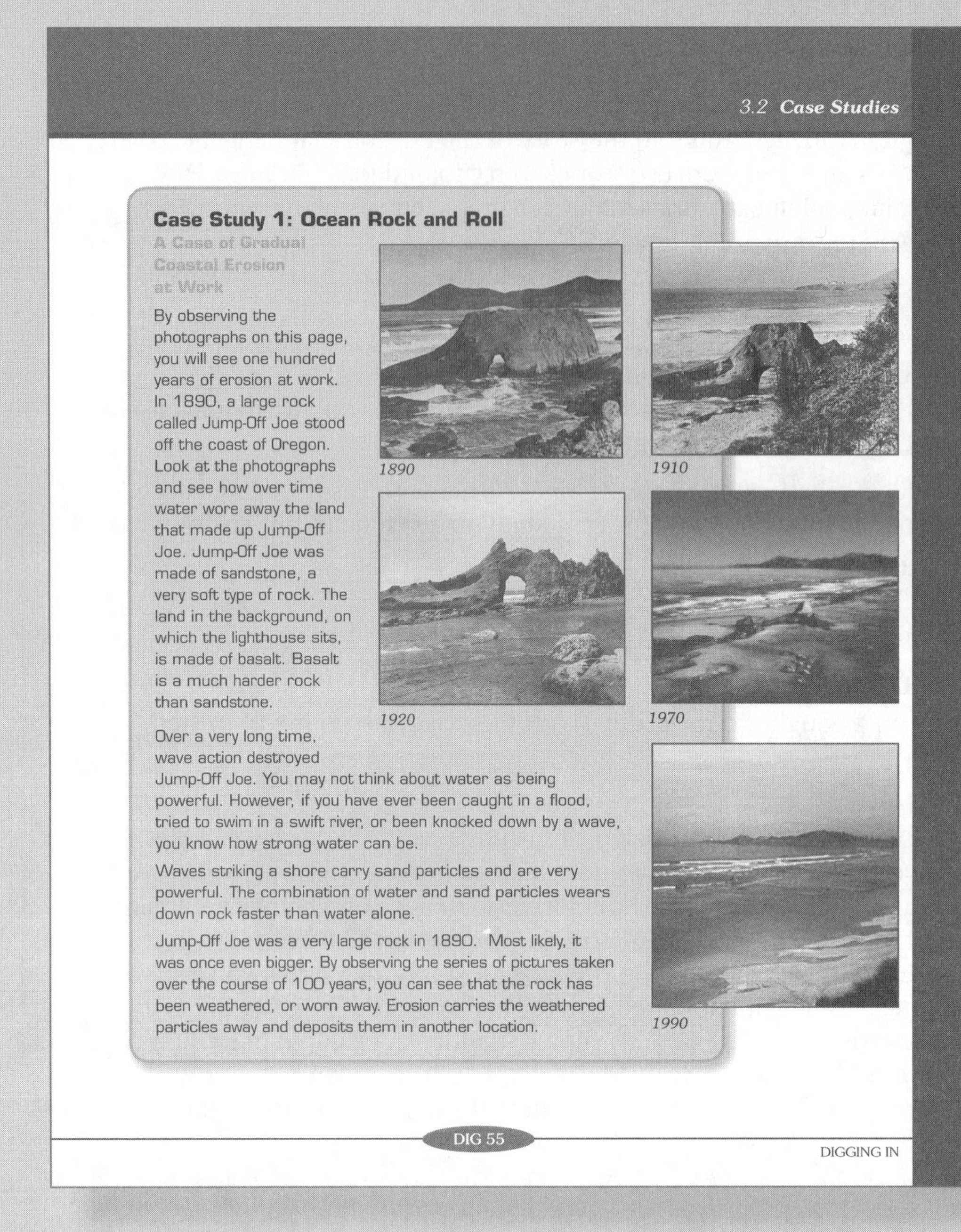

Case Study 1: Ocean Rock and Roll

A Case of Gradual Coastal Erosion at Work

By observing the photographs on this page, you will see one hundred years of erosion at work. In 1890, a large rock called Jump-Off Joe stood off the coast of Oregon. Look at the photographs and see how over time water wore away the land that made up Jump-Off Joe. Jump-Off Joe was made of sandstone, a very soft type of rock. The land in the background, on which the lighthouse sits, is made of basalt. Basalt is a much harder rock than sandstone.

1890

1910

1920

1970

1990

Over a very long time, wave action destroyed Jump-Off Joe. You may not think about water as being powerful. However, if you have ever been caught in a flood, tried to swim in a swift river, or been knocked down by a wave, you know how strong water can be.

Waves striking a shore carry sand particles and are very powerful. The combination of water and sand particles wears down rock faster than water alone.

Jump-Off Joe was a very large rock in 1890. Most likely, it was once even bigger. By observing the series of pictures taken over the course of 100 years, you can see that the rock has been weathered, or worn away. Erosion carries the weathered particles away and deposits them in another location.

When answering the questions, elicit students' ideas and encourage discussion. Have students describe their reasons for their answers. Ask students what they think conditions were like before the erosion in this case. Have them write the answer in their *Erosion Case Study* pages, under *What was happening before the erosion?* Ask them what erosion happened. Continue answering questions on the *Erosion Case Study* page until it is complete.

Erosion Case Study 3.2.1

Name: ______________________ **Date:** __________

Case name: *Case Study 1: Ocean Rock and Roll*	
Case information: time, location *1890 to 1990* *Jump off Joe on Oregon's Coast*	
The situation	
The setup: What was happening before the erosion? What kinds of things was the land being used for? *The land is on the coast, it is not used for any specific purpose. There is a lighthouse far away from it.* *The land does not appear to be used for anything in particular.*	The problem: What erosion happened, and what problems, if any, did the erosion cause? *The big rock, made of soft sandstone, known as Jump-off Joe, was destroyed.* *The erosion did not cause any known problems, it just eroded the rock almost completely.*
The erosion	
Causes: What caused the erosion? *Water waves mixed with sand caused the erosion.*	Time: How long did it take for the erosion to happen? *The story discusses erosion throughout a 100 year span.*
Solutions	
Fixes: How did they try to fix it? Who tried to fix it? *There is no mention of trying to prevent the erosion.*	Outcomes: What, if anything, happened as a result? *Not applicable.*
Conclusions: What can we learn from this case?	
About causes of erosion? *Water waves mixed with sand can erode rocks.*	Anything else? *Some students may mention that this is an example of moving water causing erosion. Some students may list examples of this type of erosion from their walk. Others might list examples with water waves. Some may say they did not see anything like this erosion on their walk because they did not see waves.*

When the class has completed the *Erosion Case Study* page, record the causes of erosion from the case study, as described in the procedure. You should keep this list out to add on to after all groups have presented their case studies.

Connect the erosion described in *Case Study 1* to examples of erosion that students observed on their walk. Some may mention that this is an example of moving water causing erosion and list examples of this type of erosion from their walk. Others might just list examples with water waves and say they did not see anything like this erosion on their walk because they did not see waves.

The Cape Hatteras Lighthouse, 1985. The ocean shoreline is just 46 m (150 ft) from the lighthouse.

The Cape Hatteras Lighthouse, 2000. The lighthouse was moved back from the ocean 884 m (2900 ft) in a massive engineering project.

Case Study 2: Where's the Beach?

Moving the Cape Hatteras Lighthouse

The Cape Hatteras Lighthouse was built in 1870. At that time, people did not fully understand the forces of erosion. So building the lighthouse 457 m (1500 ft) from the water seemed reasonable. For over a hundred years, the lighthouse warned ships away from dangerous waters in a part of the ocean called the "Graveyard of the Atlantic." However, after 129 years, the lighthouse itself was in danger. Coastal erosion had worn away about 396 m (1300 ft) of beach. The lighthouse now sat within 46 m (150 ft) of the very waters it had warned so many sailors to stay away from. In 1999, the lighthouse was moved 884 m (2900 ft) back to save it from falling into the ocean.

The Cape Hatteras Lighthouse is located on the Outer Banks of North Carolina's barrier islands. Barrier islands are found all along the eastern coast of the United States, parallel to the mainland's shoreline, and along many other shorelines around the world.

These long, sandy islands protect the mainland from the winds and pounding waves of the sea. But the islands are constantly changing, eroding in one place and building up in another. This is the result of waves, currents, winds, storms, and a rising sea level.

Seen from above, a view of the barrier islands off the Outer Banks of North Carolina.

Barrier islands are often in need of protection from the forces of erosion. After all, if they were to disappear, mainland shorelines would be defenseless against the seas. Their best defense is the dunes.

META NOTES

Case 2 discusses erosion with several contributing causes and some reasons, sometimes subtle, that the attempted solutions did not always work. Because it is a complicated case with some subtleties, this might make a good case to assign to a high level reading group.

○ Get Going

Some of the case studies are difficult and some are easy, try to select groups by reading level. Let them know that they will look at case studies with their groups and they will all present their results to the class.

Assign two groups to each case. If your class has fewer than six groups, assign two groups to each of the most difficult cases, and assign one group to each of the easier cases. Give students a time frame and have them get started.

Erosion Case Study

3.2.1

Name: ______________________ Date: ____________

<table>
<tr><td colspan="2">Case name: Case Study 2: Where's the Beach?</td></tr>
<tr><td colspan="2">Case information: time, location
1870 to 1999
Cape Hatteras, NC</td></tr>
<tr><th colspan="2">The situation</th></tr>
<tr><td>The setup: What was happening before the erosion? What kinds of things was the land being used for?
• Before erosion, there was a lighthouse about 457 m from the coast on a barrier island, which warned ships of dangerous waters. The land surrounding the lighthouse had vegetation during this time.
• This type of land is a barrier island. Barrier islands help protect the main land's coast by reducing the force of the waves.
• Barrier islands are used for recreation. Many people build homes and hotels along the beaches.</td><td>The problem: What erosion happened, and what problems, if any, did the erosion cause?
• The beachfront was eroded and a lighthouse was in danger of being destroyed. The lighthouse was now only 46 m (150 ft) away from the water.</td></tr>
<tr><th colspan="2">The erosion</th></tr>
<tr><td>Causes: What caused the erosion?
• Ocean waves (water waves mixed with sand).
• A rapid rise in the average ocean levels.
• The gradual sinking of coastal land.
• Efforts to reduce erosion that did not work, and instead increased it.
• Global warming which is expected to speed up the rise in sea level</td><td>Time: How long did it take for the erosion to happen?
• The story discusses 100 years of erosion for the barrier island, Cape Hatteras.
• The case study also describes other places where erosion occurs in coastal areas due to waves, but does not mention the time period involved for other situations.</td></tr>
<tr><th colspan="2">Solutions</th></tr>
<tr><td>Fixes: How did they try to fix it? Who tried to fix it?
To prevent the lighthouse from falling into the ocean, they decided to move it back 884 m.</td><td>Outcomes: What, if anything, happened as a result?
• By moving the lighthouse, the lighthouse will remain intact longer, but has no affect on the erosion.</td></tr>
</table>

△ Guide and Assess

Sometimes students have difficulty focusing on the task and getting started. Remind students to move quickly through the task and make sure that they understand what to do. If students have not started after about five minutes, check with pairs to make sure they understand the task and encourage students to focus on the task.

Dunes are anchored in place by the deep roots of dune plants. Most beach communities work hard to protect their dunes, asking people to stay off the dunes and not pick the dune plants. Still, barrier islands are eroding at incredible rates. Because of their beauty and recreational value, barrier islands are popular places to build homes and visit. Often, people build along these beaches without considering erosion.

Beach erosion has many causes:

- building houses and hotels near the ocean;
- a rapid rise in average ocean levels;
- the gradual sinking of coastal land;
- efforts to reduce erosion that have not worked and instead have increased erosion; and
- global warming, which will speed up the rise in sea level.

But erosion is not all bad. Without erosion, there would be no beaches, dunes, barrier islands, or bays. Bays are bodies of water found between barrier islands and the mainland. They are productive nurseries for many marine organisms.

No Perfect Solutions to Erosion Problems

Places with buildings on the edges of cliffs above beaches often have serious problems with erosion. As cliffs began to erode, as they did in the city of Miami Beach, Florida, the ocean was getting closer to the buildings. People were afraid that the buildings, including many homes, would be destroyed if the cliffs collapsed.

Many cities faced with these problems try and stop the erosion of cliffs in different ways. Several of these ways have been tried in Miami Beach, and all along the Atlantic coastline. To slow down erosion, city engineers often build barriers along beaches or into the water. Structures built include seawalls—made of concrete, steel, or wire cages filled with pebbles, or groins and jetties—different types of barriers made of rocks. They are designed to keep ocean currents from carrying away sediment and sand. Often these structures shift the movement of **sediment** or sand to other parts of the beach, causing more damage in another area.

Breakwaters, long heaps of rocks dumped parallel to the shore, reduce the strength of waves before they reach the beaches. Some people object to breakwaters because they can spoil ocean views.

sediment: solid fragments of inorganic or organic materials that come from rock and are carried and deposited by wind, water, or ice.

NOTES

Jetties (left) are similar to groins, but are used to keep sand away from shipping channels. This erosion-control method helps one area of the beach but hurts another area.

Breakwaters (left), are promising solutions to the problem of beach erosion. Some people object to the way breakwaters block views of the ocean.

Another way to restore beaches is to pump lots of sand onto them through a process called "beach nourishment." Just as food nourishes bodies, the sand, taken from deep in the ocean or construction projects, helps to build up the beaches. This process is expensive and, because erosion continues to remove the sand, beach nourishment must often be repeated after several years.

Seawalls (left), made of concrete, rock, steel, or wire cages filled with pebbles, are built in many places to slow down the effects of erosion.

NOTES

Erosion Case Study

3.2.1

Name: ______________________ Date: ____________

Case name: *Case Study 3: Landslides*	
Case information: time, location *2005* *Laguna Beach, CA*	
The situation	
The setup: What was happening before the erosion? What kinds of things was the land being used for? *Heavy rain from months earlier accumulated in the ground, reducing friction between the rocks and the underlying ground.* *The land was used for housing.*	The problem: What erosion happened, and what problems, if any, did the erosion cause? *A landslide occurred and 11 homes were destroyed.*
The erosion	
Causes: What caused the erosion? *Heavy rains that saturated the soil and reduced the force of friction between the particles of soil caused the erosion. As the soil became saturated with water, the force of friction between the soil particles was reduced, making it easier for the soil to move. The soil began to move like a fluid.*	Time: How long did it take for the erosion to happen? *The case study states that months of heavy rains occurred. No further details were provided.*
Solutions	
Fixes: How did they try to fix it? Who tried to fix it? *The case did not mention any fixes for Laguna Beach, CA. It was mentioned that vegetation helps prevent landslides by absorbing the water in the soil and by holding the soil in place.*	Outcomes: What, if anything, happened as a result? *Not applicable.*
Conclusions: What can we learn from this case?	
About causes of erosion? • *Landslides occur when the ground becomes saturated with water and no longer holds together. The water reduced the friction between the rocks and surrounding soil.* • *Sometimes the flow of water-saturated soil is on the surface and sometimes it occurs below the surface with the top layer of the soil being carried away with it.*	Anything else? • *After a landslide occurs, the rate of erosion by wind and water increases. This could be because the soil is exposed without vegetation on it. The more water that gets into the ground and rocks, the more unstable the ground becomes.* • *This is an example of moving water causing erosion. Students may list other examples.*

NOTES

Case Study 3: Landslides

A Case of Gravity and Water

A landslide is the sliding downhill of loose rocks and soil. Landslides occur when gravity pulls on rocks and soil. You can see from the picture how a landslide has left a huge scar that will quickly lead to further erosion. Landslides happen when the forces holding soil or rock together are smaller than the force of gravity pulling them down. For example, fires sometimes burn the trees and brush on steep mountains. The roots of the trees and brush help hold the soil in place. Once they are gone, the soil and rocks slide down the slope.

Once a landslide has occurred, the rate of erosion by water, gravity, and wind speeds up. Erosion leads to more erosion. Rain can get into rocks and cause them to become unstable. When unstable rock and soil get wet, they get heavier. The force of gravity pulls them more strongly. Land often moves more after a soaking rain.

As a child, you may have experienced the effects of a landslide yourself. If you try to build a sand castle with very dry sand, the sand tumbles downhill as it is pulled by gravity.

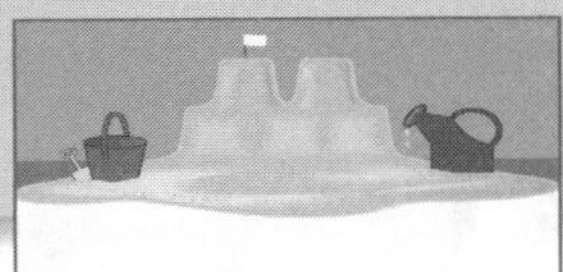

When you use moist sand, the sand particles stick together and the sand castle remains sturdy.

If too much water is added to the sand, the sand will become fluid-like, and it will flow downhill. The extra water strains the forces holding the sand together, and then gravity pulls the sand downward.

NOTES

friction: a force that resists motion.

Laguna Beach Landslide

In 2005, the Laguna Beach landslide in California destroyed at least 11 homes. Many people were evacuated. However, this landslide could have been predicted, according to many people..

The landslide occurred because heavy rain from months earlier accumulated in the ground. This wetness reduced **friction** between the rocks the homes were built on and the underlying ground. Water accumulated in layers deep beneath the surface. If the water content of soil becomes high enough, the soil will flow like a fluid. At the Laguna Beach location, the soil deep beneath the surface became fluid-like and began to flow downhill. Drier soil and rocks from upper layers and the surface were carried along on top of the flow. This caused the landslide to push and batter homes in its path, instead of flowing around them.

Landslides happen on steep slopes. Builders in Laguna Beach should have studied the conditions before building homes in the area. But, many of the houses in this area were built before current building laws were in place. There are a lot of areas in Southern California that have conditions similar to those in Laguna Beach.

The steep slopes and rainstorms in the coastal town of Laguna Beach, California, have resulted in serious landslides, as many people anticipated. After the landslide of 2005, over a thousand people were evacuated from 500 homes and much property was destroyed or severely damaged.

NOTES

Case Study 4: The Dust Bowl

Erosion Caused by Wind

The Dust Bowl occurred in the middle region of the United States, including areas of Kansas, Texas, and Oklahoma. The Dust Bowl was the name given to a 10-year period of drought that occurred in the 1930s. During this time, many people suffered great hardships, and many died.

The Dust Bowl happened because people came to the area known as the Great Plains and started plowing and farming the land. This land was not ideal for farming, but the settlers did not understand this. They did not know how to farm the plains and did not understand the effects farming could have on the land.

Before the Civil War, when settlers first passed through the Great Plains, the area between the Mississippi River and the Rocky Mountains was dry. It did not seem worth staying there, as there was no gold to be found, and the land could not be farmed. These early settlers continued on to the west coast. On old maps, they called this area "The Great American Desert."

Settlers began arriving again in the 1880s, after a period of exceptionally heavy rains. The plains were bursting with tall grass and appeared to be ideal for farming. Few people remembered how dry the plains had been just 20 years before.

People mistakenly believed that farming itself would cause more rain to fall. They also thought that building railroads and bringing in electric wires would cause more rain to fall by changing the natural electric cycles of the air. In the 1890s, there was a short drought, but soon the rains came again. It seemed like rain was normal and droughts were unusual.

In the 1930s, the drought returned, and it stayed for 10 years. The farmers had broken up the prairie soil and plowed under the native grasses. They then planted wheat. But the wheat could not survive in a drought like the grasses could. When the wheat died, its roots no longer held the soil in place. Farms turned into deserts covered with blowing sand. Huge dust storms whipped millions of tons of soil into the air. Dust storms blew soil from Kansas all the way to New York City.

Cattle were found dead in the fields with two inches of dust coating the insides of their stomachs. People coughed up clumps of dirt from the dust they had been breathing. Many of the people left the area looking for a better life somewhere else. They became known as "Okies" since many of them came from Oklahoma. A reporter, writing about one of the

drought: a long period of dry weather with very little or no rain.

NOTES

largest dust storms, called the area the "Dust Bowl," and the name stuck. The wind blew fertile topsoil away. Even today, this area has not completely recovered. Unfortunately, the Dust Bowl could have been avoided if the settlers had recalled the dry history of the area, had used different farming methods, and had not overplowed and overgrazed the land.

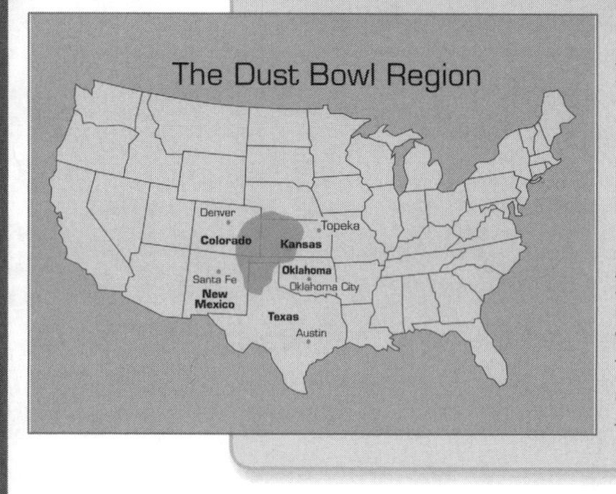

The Dust Bowl is the name given to the area of huge dust storms caused by many years of drought in the 1930s. The lack of rain, along with the plowing under of the prairie soil in the Great Plains, including Kansas, Oklahoma, and Texas, caused millions of tons of soil to blow into the air. Thousands of farms were abandoned, and many people lost their homes and suffered many illnesses from all the dust and dirt in the air. The effects of the Dust Bowl are still felt today in the Great Plains, where erosion caused by the dry winds blew away much of the fertile topsoil.

NOTES

Erosion Case Study

3.2.1

Name: ______________________ **Date:** ____________

Case name: *Case Study 4: The Dust Bowl*	
Case information: time, location *10 years, beginning in the 1930s,* *Middle region of US, including Kansas, Texas, and Oklahoma*	
The situation	
The setup: What was happening before the erosion? What kinds of things was the land being used for? *The region was mainly prairie containing natural grasses that could handle droughts, but the natural vegetation was removed when farmers moved on to the land.*	The problem: What erosion happened, and what problems, if any, did the erosion cause? *Erosion caused by wind occurred. The wind carried dry soil (sand) through the air. These dust storms contributed to the deaths of humans and other animals.*
The erosion	
Causes: What caused the erosion? *A drought of 10 years caused the land to dry up. Because the farmers had removed the natural vegetation and plowed the land, the dirt was loose and not held together by the roots of natural grasses. As the wind blew, the dry soil was picked up and carried through the air, creating dust storms.*	Time: How long did it take for the erosion to happen? *It is not clear, but the particular drought in the case study lasted for 10 years.*
Solutions	
Fixes: How did they try to fix it? Who tried to fix it? *No fixes were mentioned. It does state removing the natural vegetation such as the natural grasses, would not be helpful.*	Outcomes: What, if anything, happened as a result? *Since no fixes were used, the result was huge dust storms, dead cattle with two inches of dust lining their stomachs, and people coughing up dirt.*
Conclusions: What can we learn from this case?	
About causes of erosion? *Wind is capable of moving dirt around. It seems that it is easier for wind to move dry dirt than wet, because dry dirt is lighter.*	Anything else? *Some may mention that this is an example of moving wind causing erosion and list examples of this type of erosion from their walk. Some may say they did not see anything like this erosion on their walk.*

NOTES

Communicate: Share Your Case Study

20 min.

Students present and discuss what they learned about erosion.

Communicate

Share Your Case Study

Each of the case studies in this section presents a different example of erosion. The environment was different in each case, as were the materials involved and the factors that caused the erosion. Since each group read only one case study, it is important that the information and lessons learned are shared with the class. Use the information you recorded on your *Erosion Case Study* page to make a presentation to your class about what you learned from the case study you read. Since your classmates will be relying on you to learn about different examples of erosion and how they occurred, it is important for you to include the following information:

- the location and time of the case;
- a description of what happened;
- how long it took for erosion to happen;
- conditions that caused the erosion to happen;
- solutions to the problem and how well they worked; and
- any negative side effects caused by the solutions.

It is important that your presentation includes answers to all of the above questions. As you listen to your classmates' presentations, it is just as important that you hear and understand the answers to the same questions. If you do not understand something, or if you think presenters left out something important, ask questions. Be careful to ask your questions respectfully and not interrupt your classmates' presentations.

Reflect

Now that you have heard about several different cases of erosion, you know a lot about erosion and what causes it. You even know some things about how to manage erosion. Taking all of the cases into account, answer the questions below. Be prepared to discuss your answers in class.

1. Based on your erosion walk and the case studies you have just read, define erosion in your own words.
2. What are two forces of nature that seem to have a significant role in erosion? What effect do they have on eroding soil, sand, and other materials?

META NOTES

If you have students who you think might benefit from presenting the information visually, consider having groups create posters.

△ Guide

Guide students in preparing their presentations.

When groups have finished gathering information from their case studies, let them know they will now design their class presentations so the entire class can learn from all the studies.

Emphasize that groups' presentations should include the information specified in the bulleted list in the student text. Remind them they are responsible for informing their classmates about the case of erosion they read about.

TEACHER TALK

"Remember that we are learning from your presentation about the causes of erosion in your case study. It is important that you include all the information listed in your text in your presentation. Use these to guide you as you create your presentations."

□ Assess

Monitor students' progress as they prepare their presentations. Check that they are answering the bulleted questions from the student text. It is important that the class hears about the type of erosion, its causes, possible methods for controlling this type of erosion, and the pros and cons of trying to control it using a certain method.

Look for similarities and differences between groups analyzing the same case study. You may want to highlight these during the class discussion.

△ Guide

Have groups presenting the same case study present one after another.

Engage the class in discussion and emphasize that they should direct questions to the presenting group. Model the kinds of questions you expect students to ask.

TEACHER TALK

"I'm not sure I understand what led to the erosion. Can you explain what some of the reasons for this erosion were?

Can you explain how this solution to erosion led to the side effects you mentioned?

I'm not clear on what the conditions were before the erosion occurred. Could you clarify?"

If students are not using posters in their presentations, record important information from their presentations on the board.

Discuss the results of two groups presenting the same case study after they have presented.

After two groups with the same case study have presented, discuss the similarities and differences between their answers. If there are differences, ask the groups why they think those differences occurred. Have groups present their reasoning behind their responses. If differences in responses still exist, go back to the case and consider as a class each group's response and discuss it. Note that these are things that may go on the class *Project Board* later in this section.

◇ Evaluate

Make sure that each group has addressed the bulleted items in the list in the student text. It is important that students understand the cause of erosion for each case study.

△ Guide

After all groups have presented, hold a discussion on the causes of erosion and possible ways of controlling them, using the list you recorded from the presentations as the focal point. Ask if there is anything else to be added. Ask what was similar and what was different from the case studies. They may be able to point out that the primary factor causing erosion in Cases 1, 2, and 3 is water.

Reflect

15 min.

Students participate in a discussion of the Reflect *questions.*

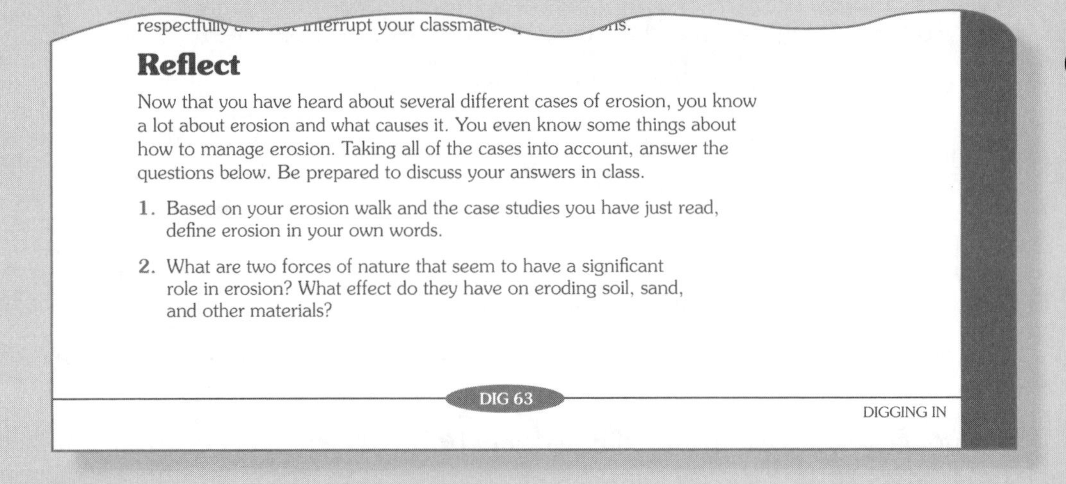

respectfully ... interrupt your classmates ... ons.

Reflect

Now that you have heard about several different cases of erosion, you know a lot about erosion and what causes it. You even know some things about how to manage erosion. Taking all of the cases into account, answer the questions below. Be prepared to discuss your answers in class.

1. Based on your erosion walk and the case studies you have just read, define erosion in your own words.
2. What are two forces of nature that seem to have a significant role in erosion? What effect do they have on eroding soil, sand, and other materials?

DIG 63 DIGGING IN

Have students answer the *Reflect* questions individually. Let students know how much time they have to answer the questions and that they will be discussing their anwers with the class.

△ Guide and Assess

Lead a class discussion of students' answers. Listen for the following in students' responses:

1. Students' answers should indicate that erosion is the displacement of material by water, ice, or wind.

2. Students should mention friction and gravity as two forces that affect erosion. Students should give their reasons for picking each force, how they think it affects erosion, and examples of each from the case studies. Assist students in understanding how these forces affect erosion. Friction tends to hold material in place, and when friction is lessened by the introduction of water, erosion can increase by making the material easier to displace. This is apparent, in the case of the landslide at Laguna Beach. Gravity is involved in many ways. The force of gravity pulls all things near Earth toward the center of Earth. It is the cause for water and ice flowing downward, and water-laden material.

NOTE: The *Reflect* questions in the student text are divided by an *Erosion and Weathering* textbox. Make sure students understand these are a part of the segment and they need to answer them with the others.

NOTES

Erosion and Weathering

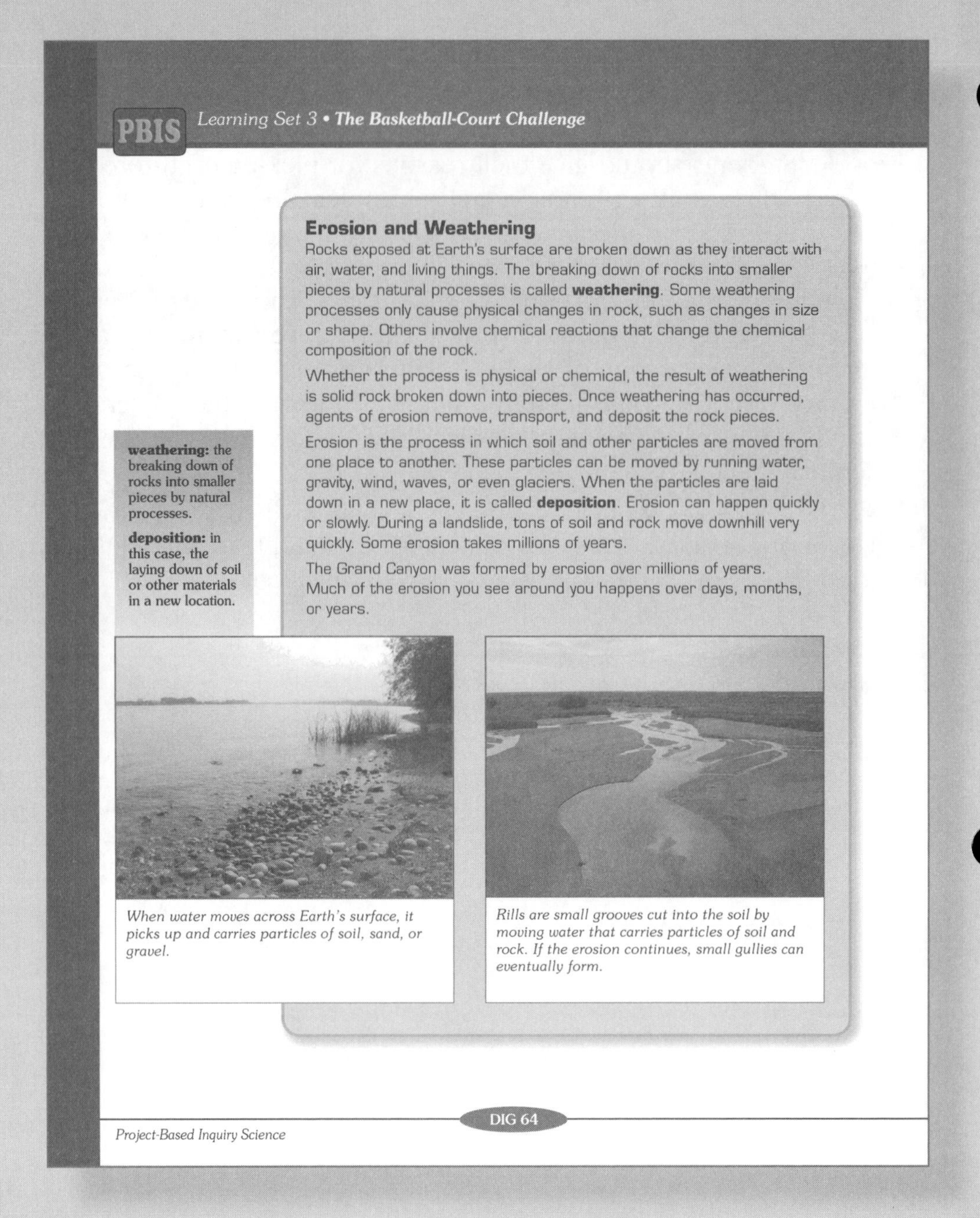

Erosion and Weathering

Rocks exposed at Earth's surface are broken down as they interact with air, water, and living things. The breaking down of rocks into smaller pieces by natural processes is called **weathering**. Some weathering processes only cause physical changes in rock, such as changes in size or shape. Others involve chemical reactions that change the chemical composition of the rock.

Whether the process is physical or chemical, the result of weathering is solid rock broken down into pieces. Once weathering has occurred, agents of erosion remove, transport, and deposit the rock pieces.

Erosion is the process in which soil and other particles are moved from one place to another. These particles can be moved by running water, gravity, wind, waves, or even glaciers. When the particles are laid down in a new place, it is called **deposition**. Erosion can happen quickly or slowly. During a landslide, tons of soil and rock move downhill very quickly. Some erosion takes millions of years.

The Grand Canyon was formed by erosion over millions of years. Much of the erosion you see around you happens over days, months, or years.

weathering: the breaking down of rocks into smaller pieces by natural processes.

deposition: in this case, the laying down of soil or other materials in a new location.

When water moves across Earth's surface, it picks up and carries particles of soil, sand, or gravel.

Rills are small grooves cut into the soil by moving water that carries particles of soil and rock. If the erosion continues, small gullies can eventually form.

△ Guide

Discuss erosion and weathering with the class. Weathering breaks rocks down into smaller pieces. There can be many causes of weathering. When water freezes in the cracks in a rock, it may break the rock apart. When a plant grows in the cleft of a rock, its roots may break the rock apart.

Erosion moves particles, such as soil. Like weathering, erosion can be caused by many different things, some of which students have already seen.

It is easy to confuse erosion and weathering. Ask students what the difference between these two processes is and help them if they have trouble understanding the distinction.

3. What are some human activities that cause erosion?
4. Describe some ways erosion can be controlled. What are some negative side effects of trying to control erosion?

Update the *Project Board*

You have learned about erosion from direct observation during your *erosion walk* and by reading several case studies. Now you will put these ideas about erosion into your *Project Board*, so everyone in the class can think about the next steps in the challenge.

As a class, review and discuss the *Project Board*. You can update or record new facts in the *What are we learning?* and *What is our evidence?* columns. When you record what you are learning in the third column, you will be answering some questions in the *What do we need to investigate?* column. You will describe what you learned from the case studies you just read.

But you cannot just write what you learned without providing the evidence for your conclusions. Evidence is necessary to answer scientific questions. You will fill in the evidence column based on the descriptions and analyses of large-scale erosion that you found in the case studies. You may use the text in this book to help you write about the science you have learned. However, make sure you put it into your own words. The class will fill in the large *Project Board*. Make sure to record the same information on your own *Project Board*.

As you read the different cases in this section, you may have thought of some new questions or ideas that you are not sure about. These can be added to the *What do we need to investigate?* column. In order to meet the challenge, you will need to learn more about erosion and the factors that affect erosion. These are some of the things you still need to investigate.

Update the *Project Board*

15 min.

Students are introduce to columns 3 and 4 of the Project Board.

META NOTES

In science it is very important that claims be supported by evidence. The evidence students use should be based on their observations, case studies, or science knowledge.

△ Guide

As students discuss their answers to the *Reflect* questions, ask them how what they have learned relates to the *Basketball-Court Challenge.* What causes erosion and how can it be controlled? Display the class *Project Board* and remind students that the *Project Board* is a way to organize their ideas and questions. It helps them see how their ideas change as they gather more information.

Remind students what they have already recorded, and remind them of what the five columns are: *What do we think we know? What do we need to investigate? What are we learning? What is our evidence?* and *What does it mean for the challenge or question?*

Tell students they will be updating their *Project Board* with what they have learned from the cases they just read and that this information will go primarily into the third (*What are we learning?*) and fourth (*What is our evidence?*) columns. Point out that they may also have new questions they want to investigate that can be added to Column 2.

First model how to put information in columns 3 and 4 by emphasizing that Columns 3 and 4 are directly linked. Remember to draw arrows directly linking the claim and evidence on the *Project Board*. You may want to use the example below

META NOTES

If claims cannot be supported by evidence, then an investigative question should be developed and placed in Column 2.

Column 3 (claim): Waves from the sea can wear down large rocks completely.

Column 4 (evidence): We learned that the large sandstone rock Jump-Off Joe was worn down completely by erosion from 1890 to 1990.

Begin the discussion by asking students what else they know based on what they read today and what they learned from the erosion walk. Update the *Project Board* with the class. Remember to draw arrows indicating the flow of ideas and to date the entries.

TEACHER TALK

"What did you learn from your case studies? What is the evidence that supports that idea?

What did you learn about erosion that is not on the *Project Board?* What should we write as evidence?

What did you learn about how erosion can be controlled? What should we write as evidence?"

◇ Evaluate

Make sure students include claims about how erosion is caused by wind and water in the form of waves or rain, and gravity in column 3. They should have questions in Column 2 about how these cause erosion and what influences how erosion happens. There should also be evidence from the case studies in Column 4.

NOTES

What's the Point?

When scientists and engineers are confronted with a problem, others have often had to deal with the same problem before. It is helpful to investigate and read about such similar cases. Case studies can help you understand a problem. They can also show you how others have attempted to solve similar problems and how well their solutions worked. If the solution worked, then you can try to improve upon it and make it work better. If it did not work, then you know that you have to try a different approach to solving the problem.

You had the opportunity to read four case studies. Each case showed how a different force in nature can cause erosion. You saw examples of erosion by rain, waves, gravity, and wind. Later on you will read more about some of the different methods used to control erosion. Some have been successful while others have not. This information will be very valuable to you as you work to achieve success at the *Basketball-Court Challenge*.

Beach erosion, like other types of erosion, results from a combination of factors: construction near shorelines, a rapid rise in ocean levels due to global warming, and the gradual sinking of coastal lands.

Gravity pulls soil, rocks, and other particles down slopes or hillsides, causing erosion.

Teacher Reflection Questions

- Which case studies were most difficult for students? How can you help students understand this content?
- How were you able to relate the cases to the *Basketball-Court Challenge?* What can you do to help students make connections between the cases and their challenge?
- How were you able to keep students engaged in presentations? What types of questions did students ask of presenting groups? How can you encourage their questioning?

SECTION 3.3 INTRODUCTION

3.3 Investigate

Investigating Factors that Affect Erosion

2 class periods ▶

A class period is considered to be one 40 to 50 minute class.

Overview

Students run investigations to determine how slope and the composition of Earth materials in the ground affect erosion. Half of the class investigates what happens when hills of sand with different slopes are saturated with water. The other half investigates how different materials in a hill erode when water runs over the hill. Groups present their results and interpretations on posters during their first *Investigation Expo*. Then students are introduced to fair tests and think about how they worked with different kinds of variables (independent, dependent, and control). Like scientists, students share and learn from each other's observations and ideas about erosion.

Targeted Concepts, Skills, and Nature of Science	Performance Expectations
Scientists often work together and then share their findings. Sharing findings makes new information available and helps scientists refine their ideas and build on others' ideas. When another person's or group's idea is used, credit needs to be given.	Students should share their results and claims with the class. They should use what they learned from each other to create explanations (in *Section 3.4*) .
Scientists must keep clear, accurate, and descriptive records of what they do so they can share their work with others and consider what they did, why they did it, and what they want to do next.	Students should record their observations on *Particle Size and Erosion* and *Slope and Erosion* pages, and use their observations to make their interpretations and prepare their presentations.
Identifying factors that could affect the results of an investigation is an important part of planning scientific research.	Students should recognize which variables they were manipulating in this investigation.

Targeted Concepts, Skills, and Nature of Science	Performance Expectations
Scientific investigations and measurements are considered reliable if the results are repeatable by other scientists using the same procedures.	Students should recognize that groups doing the same procedures should get consistent results.
In a fair test only the manipulated (independent) variable and the responding (dependent) variable change. All other variables are held constant.	Students should recognize the variables that they held constant in their investigations.
Scientists make claims (conclusions) based on evidence (trends in data) from reliable investigations.	Students should look for trends in their data and make claims based on the trends.

Materials	
For *Investigation 1* (Only half the class will be conducting this investigation.)	
1 cup per group	Pre-wetted mixture of sand, potting soil, gravel, native soil
1 per group	Plastic drop cloth Stream table tray Small cup with water Ruler Presentation materials
1 per half class	Measuring cup
As needed, per group	Books
1 per student	Disposable gloves *Particle Size and Erosion* page
For *Investigation 2* (Only half the class will be conducting this investigation.)	
2 cups per group	Pre-wetted, fine sand

Materials	
For *Investigation 2* (Only half the class will be conducting this investigation.)	
1 per group	Plastic drop cloth Stream table tray Small cup, with water Spray bottle, filled with water Measuring cup Ruler Presentation materials
1 per half class	Measuring cup
As needed, per group	Books
1 per student	Disposable gloves Slope and Erosion page
1 per classroom	Stopwatch Paper towel roll

NOTES

Activity Setup and Preparation

These two investigations contain water and numerous Earth materials including dirt, which has the potential to create a mess. Consider how you want to organize your classroom, the materials, and how you will monitor students during the investigations.

Students will be using water and Earth materials to see what happens when water flows over the Earth materials. During and after the class, students will need to clean up these very messy materials, rinsing out cups and washing out bins for the next class. Decide how you want students to clean up the bins and the materials. Students will need to empty water from the bins or the cups they use to perform the investigation. This excess water should be poured into a large bucket. You will need to empty the bucket, as it fills up, in a proper place or receptacle. This water will contain sediment that should not be poured down a sink drain. Use large bins to reclaim Earth materials after the experiments.

In this section, students will present posters to the class using *Investigation Expos*. This is the first time students give presentations using *Investigation Expos*, so they will need guidance. Before class, prepare a sample poster showing all of the information that should go in to students' presentations. Students' posters will need to include their investigation question, the predictions they made, their procedure and what makes it a fair test, their results and their confidence in their results, and their interpretation of their results. The poster you make for the class might look like the following.

Investigation Question: What is the relationship between particle size and erosion? (Investigation 1)	
Predictions: ***What materials did you think would be most easily carried by water? Why did you think that?***	
Your procedure and what makes it a fair test: ***The procedure is given to students, but they should note anything they had to do that was not specified in their books.***	**Your results and how good you think they are:** ***Students should put the results that they recorded on their posters.***
Interpretation of your results: ***What trends did you identify in your results? What claim can you make based on the trends you identified?***	

You might also suggest that students use your poster as a template. You could make a second poster or handout with spaces for all of the information students need to include in their posters.

Homework Options

Reflection

- **Science Process:** You want to compare how granite is eroded by a river to how sandstone is eroded by a river. To test this you build a model of a river flowing over granite and sandstone. What are the manipulated and responding variables? What are some of the variables you will control? *(Students should recognize that the manipulated variable is the type of Earth (or river-bed material) and the responding variable is how much material is displaced. Some control variables are water speed, water volume, and water temperature.)*
- **Science Content:** Based on your results, which of the materials that the class studied do you think is the most susceptible to erosion by itself? The case studies your class read may have shown a more complicated picture. What are some factors that you saw in the case studies that might change how different materials erode?
(Students should recognize one striking lesson from the case studies is that grass and other vegetation very effectively protect soil from erosion.)

Preparation for 3.4

- **Science Content:** How do you think you can use your results from the investigations done in this section and from the case studies your class read to find a solution to the *Basketball-Court Challenge? (Students should make connections from the cases and their investigations to the challenge. Ideas of what to do with sloped and saturated land will be important. Students should make connections to the case study about landslides.)*

NOTES

SECTION 3.3 IMPLEMENTATION

◀ *2 class periods**

3.3 Investigate

Investigating Factors that Affect Erosion

On your erosion walk and while you were reading the cases, you may have noticed that the type of soil or other Earth materials can make a difference in how and when erosion occurs. In this section, you will investigate several different types of soil and materials to see how water and gravity affect their erosion.

Your class will complete two investigations. One half of your class will investigate the relationship between particle size and erosion. The other half will investigate the relationship between steepness of slope and erosion. Each group will collaborate to interpret their observations and then share their findings with the class. In this way, you will be able to learn from one another.

Be a Scientist

Variables and Designing Experiments

When you investigate a **phenomenon**, you want to learn about the factors that influence it. In science, these factors are called **variables**. The point of most experiments is to understand how a variable affects the phenomenon you are investigating. The phenomenon you are studying is erosion. There are many variables that affect erosion; you will focus on how the type of material and slope of the land affect erosion.

phenomenon: an event or detail that can be observed.

variable: a single factor that is tested in an experiment.

Investigation 1: What Is the Relationship Between Particle Size and Erosion?

You saw in the case studies you read that water is a very powerful agent of erosion. Many of the examples of erosion you saw on your erosion walk were probably caused by water. As water runs over Earth's surface, it picks up and carries away particles of soil and other materials. Some particles

DIG 67 DIGGING IN

3.3 Investigate

Investigating Factors that Affect Erosion

35 min.

Sudents are introduced to the investigations.

META NOTES

It is important for the entire class to understand each investigation. The class will be responsible for understanding the data from both investigations even though each group will only conduct one investigation.

Emphasize that students write down the description of procedures you are giving them because some of what you will be telling them is different from their books.

△ Guide

Begin by letting students know they will investigate how different materials erode when exposed to water and how the terrain affects erosion. Let students know that half of the class will investigate how different materials erode when water goes through them, and half the class will investigate how saturated hills of these materials erode from water flowing over them. Go over each investigation with the class, demonstrating the setup for each investigation and explaining the science involved.

As you go through the investigations, explain to students that gravity pulls the water down toward the center of Earth, and causes the water to move straight down (as in *Investigation 1*), or down a slope (as in *Investigation 2*). Gravity does not pull down on it differently, but the shape and the saturation of the material is different.

NOTES

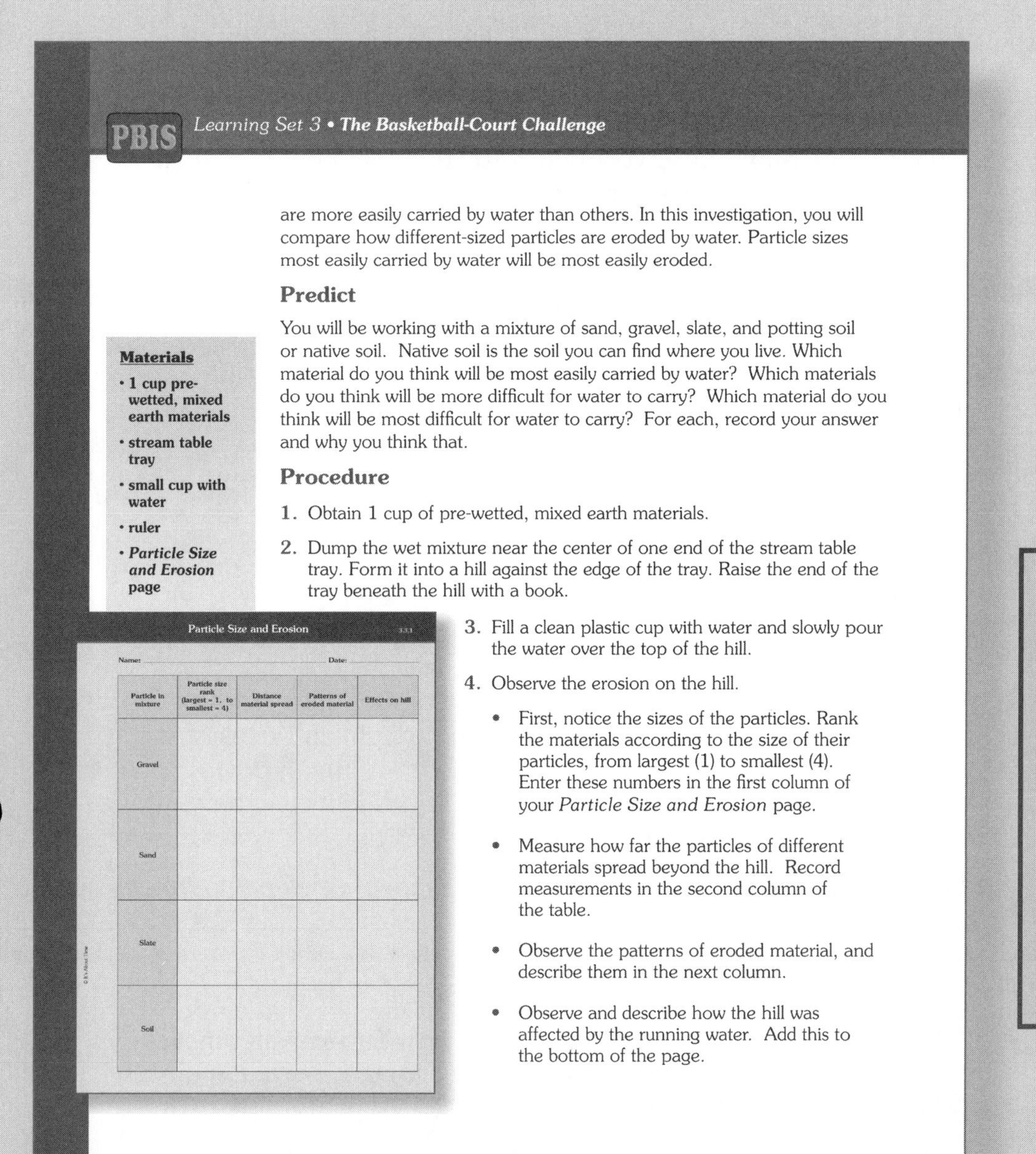

are more easily carried by water than others. In this investigation, you will compare how different-sized particles are eroded by water. Particle sizes most easily carried by water will be most easily eroded.

Predict

You will be working with a mixture of sand, gravel, slate, and potting soil or native soil. Native soil is the soil you can find where you live. Which material do you think will be most easily carried by water? Which materials do you think will be more difficult for water to carry? Which material do you think will be most difficult for water to carry? For each, record your answer and why you think that.

Materials

- **1 cup pre-wetted, mixed earth materials**
- **stream table tray**
- **small cup with water**
- **ruler**
- ***Particle Size and Erosion* page**

Procedure

1. Obtain 1 cup of pre-wetted, mixed earth materials.
2. Dump the wet mixture near the center of one end of the stream table tray. Form it into a hill against the edge of the tray. Raise the end of the tray beneath the hill with a book.
3. Fill a clean plastic cup with water and slowly pour the water over the top of the hill.
4. Observe the erosion on the hill.
   - First, notice the sizes of the particles. Rank the materials according to the size of their particles, from largest (1) to smallest (4). Enter these numbers in the first column of your *Particle Size and Erosion* page.
   - Measure how far the particles of different materials spread beyond the hill. Record measurements in the second column of the table.
   - Observe the patterns of eroded material, and describe them in the next column.
   - Observe and describe how the hill was affected by the running water. Add this to the bottom of the page.

Particle Size and Erosion 3.3.1

Name: Date:

Particle in mixture	Particle size rank (largest = 1, to smallest = 4)	Distance material spread	Patterns of eroded material	Effects on hill
Gravel				
Sand				
Slate				
Soil				

META NOTES

Students may think that the pull of gravity exerted by Earth on an object is different when the object is on a flat ground or a sloped ground. There is no difference in the pull of gravity on the object. However, the resulting motion that may occur is different due to the different topography or shape of the surface.

△ Guide and Discuss Investigation 1

Briefly discuss the procedure for *Investigation 1.* Let students know they will observe the runoff from a hill containing different materials to determine which materials erode most easily. They first make a hill by dumping a cup of a mixture of sand, gravel, and potting soil (or native soil) in one end of a stream table tray and inclining the tray by putting a book under it. They slowly pour a cup of clean water over the top of the hill and examine and analyze the erosion on the hill closely following Step 4 of the procedure. They rank the size of the eroded particles and record their ranking on their *Particle Size and Erosion* page. Have students measure how far particles of each type of material spread from the hill and record their measurements. Make sure they observe and record the patterns of the eroded materials and observe and record changes in the hill due to erosion. They should do three trials, draining the water, reshaping the hill, and getting fresh water each time.

Ask students why they should run three trials. Students should describe how this assists in determining how reliable the results are. If the three trials are basically the same, then it is likely that they have run their procedure consistently and that their results are reliable. Encourage connections with what they learned about consistent procedures and reliable results from *Learning Set 2.*

META NOTES

When doing the investigations, you might want to have students try to filter the material out to measure the mass or volume of material that passed through the cup with the water. It will be hard to measure the amount of fluid that passes through because some will be trapped in the filter and some will evaporate. By measuring the volume before filtration and the volume of material after filtration, they can figure out how much water flowed through and how much was absorbed by the material in the cup.

TEACHER TALK

"You will be running three trials for this experiment. Why? What is the benefit of running three trials?"

Emphasize that in this experiment the force of the running water dislodges some particles in the hill. Depending on the weight and density of the particles, the force of the running water may move them downstream, and the water may carry them away from the hill. Students will be observing how easily different particles are dislodged and how the water moves them.

Point out that the displacement of material students observe in this investigation is erosion.

TEACHER TALK

"For this investigation, you will measure how far the water moves particles of different materials, and observe how the shape of the hill changed and what kind of patterns formed.

When Earth materials are displaced by moving water, we call this erosion. This is the event you are investigating. Scientists call the event they are investigating the phenomenon."

Ask students what things change in this investigation. Students should state that the position of the particles in the hill, the shape of the hill, and the location of the water change. Point out that these are the variables of the experiment. Define variables as things that vary or change.

5. Drain the water from the soil, form the soil back into a hill against the edge of the tray, and repeat Steps 3 and 4. Do a total of 3 trials. Record the results for each trial.

Investigation 2: What Is the Relationship Between the Slope of the Land and Erosion?

The Laguna Beach case showed that water and gravity can work together to cause erosion. In Laguna Beach, there was so much rain that a cliff's soil became thoroughly soaked with water. The water made the cliff so heavy and loosened the bonds between the particles of soil so much that the force of gravity caused whole parts of the cliff to come loose and fall downhill. This is called a landslide.

Water and gravity also work together in smaller ways to cause erosion. On your erosion walk, you may have seen examples of soil or rocks that looked like they had rolled down a hill. In this investigation, you will experience how water and gravity together affect the way earth materials move. You will construct hills on different slopes and identify how each responds to water falling on it.

Slope and Erosion

Names: ______ Date: ______

Slope of tray	Distance material spread	Patterns of eroded material	Effects on hill	Other observations
No slope				
Gentle slope				
Steep slope				

Materials

- 2 cups pre-wetted, fine sand
- filled spray bottle
- 1 stream table tray
- small cup with water
- ruler
- *Slope and Erosion page*

Predict

You will be working with hills of sand at different slopes to see how each would be affected by rainfall. On which slope do you think materials will travel the farthest when water falls on it? On which slope do you think materials will travel the least far? Why?

Procedure

1. Obtain two cups of pre-wetted, fine sand.
2. Dump the wet sand near the center of one end of the stream table tray. Form it into a hill against the edge of the tray. Leave the tray lying flat on your desk (no additional slope).

△ Guide and Discuss Investigation 2

Discuss the procedure for *Investigation 2.* Let students know they will observe how sand soaked with water erodes under the influence of gravity. First, students make a hill by dumping two cups of wet sand in a stream table tray and shaping the sand into a hill against the edge of the tray. Then they spray water on the hill repeatedly, and observe movements of sand particles on the hill. They continue spraying water on the hill until the sand no longer moves and observe the erosion. They repeat the procedure with the hill elevated, raising the end of the tray about 5 cm. This allows students to observe how wet sand in a steeper slope erodes under the influence of gravity. After recording their observations, students repeat the process a third time with the hill elevated more, raising the end of the tray about 10 cm. This gives them a slope that is steeper. Emphasize that it is important to record all of their observations in their *Slope and Erosion* pages every time they run the procedure.

TEACHER TALK

"For *Investigation 2*, you will construct a hill of sand and soak the hill. Pay close attention to how the sand spreads as the hill absorbs more water. As you repeat the procedure with different slopes, observe how this changes the way the sand spreads."

META NOTES

We do not tell students that they have to try to always pour at the same rate and from the same height each time during both investigations. Use this as a way to assess if students understand the need for keeping these variables the same and running their procedures consistently.

Remind students when scientists study how something happens, they say they are studying a phenomenon. Ask what phenomenon they are studying in *Investigation 2*. Guide them to understand they are observing how gravity and water cause sand to erode.

Remind students the variables in an experiment are the things that change. Ask what the variables in *Investigation 2* are. They should recognize that the amount of water in the sand, the slope of the sand, and the position of the sand in the hill change.

△ Guide

Let students know that before running their investigations, they will make predictions. Groups running *Investigation 1* will predict which materials will be easiest and which hardest for water to carry. Groups running *Investigation 2* will predict which slope the sand will erode most and on which it will erode least. They should record their predictions and record the reasons for their predictions. Ask students questions like, Why do you think one material will erode faster than another? Why do you think a material will spread out more than others when water is run over it?

⬡ Get Going

Assign each group to an investigation, distribute materials (including the *Particle Size and Erosion* pages and the *Slope and Erosion* pages), let groups know how much time they have (about 20 minutes), and have groups make predictions.

Once they have recorded their predictions, they should run the procedures. Emphasize that all groups should run three trials; groups running *Investigation 2* will run three different trials and groups running *Investigation 1* will run identical trials.

△ Guide

As groups run the procedures, check to see what kinds of difficulties they are having.

- Are students who are running *Investigation 1* having trouble differentiating the materials in their hill or judging how far the different materials have spread? Are they using fresh materials for each trial? Are they pouring water onto the hill the same way each time?
- Are students who are running *Investigation 2* having trouble judging how the sand spreads in each trial? Are they keeping the elevation the same during each trial and not letting the tray slip?

□ Assess

Note whether groups running the same investigation are getting similar results. If they get widely varied results, try to help them see how they can follow the procedure more carefully.

In *Investigation 1* different groups may have poured the water differently, so their results may be different. But they should get similar trends.

Emphasize that students need to record their results as they run their investigations.

Students running *Investigation 1* will probably rank the materials as follows: 1) slate, 2) gravel, 3) sand, and 4) soil. They are likely to observe the most displacement with soil, followed by sand. Smaller gravel pebbles may be displaced. The slate is not likely to be displaced much. Students should begin to find a trend that larger particles are not displaced as much as smaller particles. There will be a lot of variation in their observations of patterns and the effects on the hill. They may observe gullies in the sand and soil.

Students running *Investigation 2* should observe the steeper the slope of the wet sand is, the more the sand is eroded. There may be a lot of variation in the patterns they observe.

3. Spray water gently on the top of the hill about 50 times. Try to spray the water on the hill the same way that rain would fall. During spraying, notice and record any movements of the sand particles on or around the hill in the *Patterns of eroded material* column of your *Slope and Erosion* page.
4. Continue spraying until the spreading of the hill slows or stops (about another 50 sprays).
5. Observe the erosion of the hill. Observe how far the particles of sand spread beyond the hill. Observe the patterns of eroded material. Observe how the hill was affected by the running water. Record your observations on your *Slope and Erosion* page.
6. Drain the water, form the soil back into a hill against the edge of the tray, and repeat the procedure, raising the end of the tray beneath the hill about 5 cm (gentle slope) in Step 2.
7. Drain the water, form the soil back into a hill against the edge of the tray, and repeat the procedure again. This time, raise the end of the tray beneath the hill about 10 cm (steep slope) in Step 2.

interpret: to find the meaning of something.

trend: a pattern or a tendency.

claim: in this case, a statement about what a trend means.

Analyze Your Results

Finding Trends and Making Claims

Your class has now collected data about how different materials are affected by erosion and how slope affects erosion. It is now time to **interpret** those results. To interpret means to figure out what something means. Interpreting results of an experiment means identifying what happens as a result of changing a variable. What happened to each of the different types of materials when the mixture was eroded by running water? How did the movement of particles of different materials compare? What spreading patterns could be observed? How did changing the steepness of the hill affect the rate of erosion? How did the steepness of the hill affect how fast water flowed downhill? How did slope affect the distance the hill spread and the amount of sand carried to the bottom of the tray?

You will do two things to interpret your results. First, you will identify **trends** in your results. Then you will state a **claim** based on those trends. A trend is a pattern that you can see over several trails. A claim is your statement about what those trends mean. For example, you varied the slope to see if it affected the rate of erosion. Your data may have shown

Analyze Your Results

10 min.

Students have collected their data, and now make a claim based on trends in their data.

○ Engage

Let students know that they will now figure out what their data mean and how their data apply to the challenge. Let students know it is important to support their ideas with reasons.

△ Guide

The groups running *Investigation 1* should compare the results of their trials and look for patterns. Did the three trials provide consistent results? This will give students an idea of how trustworthy their trials are. When they compare the particle size of the different materials with the distance the materials spread, do they see any trends?

The groups running *Investigation 2* should compare the slope of the tray to the distance the sand spread. Do they see any trends?

Explain how to make a claim based on a trend in their data.

⬡ Get Going

Tell students that the claim their group decides on should answer their investigation question and be based on their data.

You may want to ask students to write what their claim might imply for their challenge.

Remind students that they will be presenting their results and claims to the class.

□ Assess

While groups are working, listen for their ideas about trends and claims and note any topics that should be discussed when groups communicate their results.

NOTES

....

....

....

....

....

META NOTES

You could have the groups who ran *Investigation 1* make bar graphs of the particle size of each material versus the distance particles of each material spread. They should graph the data in order of the least distance to the greatest distance to show the trend.

You could also have the groups who ran *Investigation 2* make bar graphs. These groups could graph the slope of the sand hill versus the distance sand particles spread.

META NOTES

Students often confuse evidence and opinion when supporting their ideas. Scientific claims, are statements based on trends in a set of observations. These claims are based only on the data and do not contain opinions. If students bring in an opinion, guide them back to their data.

Communicate Your Results: *Investigation Expo*

40 min.

Student participate in their first Investigation Expo *to share their experimental results, interpretations, and claims on posters.*

that the finer particles in the mixture eroded faster than the larger particles over all the trials. This is a trend. Your claim would be the statement: "Finer particles erode faster than larger particles."

Every time a scientist makes a claim, other scientists look for the evidence the scientist has for that claim. One kind of evidence is data collected in an experiment and the trends in that data. You will spend a lot of time in PBIS Units making claims and supporting them with evidence. You will be learning more about that later. For now, make sure that the data you collected matches your claim.

Make sure to record the trends you have identified on your *Particle Size and Erosion* and *Slope and Erosion* pages. Also include any claims you think you can make so you can share them with your classmates.

Communicate Your Results

Investigation Expo

You will now share with the class what you have found in an ***Investigation Expo***. Remember, no groups in the class did both investigations. Therefore, others will need your results to complete the challenge. Read the box introducing *Investigation Expos* before moving on to make yourself familiar with what you will be doing in this activity.

Be a Scientist

Introducing an *Investigation Expo*

An *Investigation Expo* is like other presentations you have done, but specially designed to help you present results of an investigation. You will include your procedure, results, and interpretations of results.

Scientists present results of investigations to other scientists which lets the other scientists ask questions and build on what was learned. Scientists may present results by making posters and setting them up in large rooms at meetings with other scientists and their posters. They also give presentations about their investigations and results in front of large audiences of other scientists. Their presentations usually include visuals (pictures), showing all the important parts of their procedures and results.

To prepare for an Investigation Expo, you will usually make a poster that includes the same items that scientists' do.

***Investigation Expo*:** presentation of the procedure, results, and interpretations of results of an investigation.

META NOTES

Do not expect groups to have a lot of skill in presenting their results—students' skills will improve with practice and time. It is important for the group and the class to notice errors in the experiment (e.g., inconsistent procedures, too few trials, including outliers). The questions for the *Investigation Expo* will help to focus students on these points. It is important that the students lead the discussion as much as possible.

Describe an *Investigation Expo*. It is similar to other presentations, but it is designed for sharing information about investigations. Explain that there are two parts—poster presentations and discussions to share their procedure, results, and interpretations. Emphasize that the student text has all this information and that they should refer to the student text while preparing for the *Investigation Expo*.

These include:

- questions you were trying to answer in your investigation;
- your prediction;
- your procedure and what makes it a **fair test**;
- your results and how confident you are about them;
- your interpretation of the results (conclusions).

If you think the test you ran wasn't as fair as you had planned, report on how you would change your procedure if you had a chance to run the investigation again.

As you look at the posters and listen to other groups present their work, start with the groups that did the same investigation you did. Notice the similarities and differences in what they found and in their conclusions. If another group got different results, try to decide whose results are more accurate, yours or the others. If another group had different conclusions from yours, decide whether or not you agree with their conclusions and why.

When you look at the posters and hear the presentations of the groups that did the other investigation, make sure you get answers to all of these questions:

- What was the group trying to find out?
- What variables did they control as they did their procedure?
- Is their data consistent?
- Did they run their procedure the same way every time?
- What did they learn?
- What conclusions to their results suggest?
- Do you trust their results? Why or why not?

During the presentations, make sure you understand the procedure each group followed and that you agree with each group's conclusions. If you do not hear answers to all the questions, if the answers are not clear, or if you think a group made a mistake, ask questions. Be sure to ask your questions respectfully.

fair test: things being compared are tested under the same conditions, and the test matches the question being asked. All variables, aside from the variable you are investigating, are kept the same.

TEACHER TALK

"Throughout this class you will be using *Investigation Expos* to share ideas. During an *Investigation Expo,* you behave like scientists sharing results and their ideas. After we discuss everyone's ideas you will have a chance to reflect on what you have heard and refine your ideas."

Point out the bulleted list and emphasize that students will need to be able to inform the class of their investigative question, their predictions, their procedure, and what makes the procedure a fair test (to be discussed next), their results and how confident they are of them, and their interpretations and their confidence in their interpretations. Emphasize that they should be able to inform the class of their claims. Remind students that a claim is a statement of what a trend means or a conclusion of an experiment.

Explain a fair test. Fair tests compare things under the same conditions. For an experiment to be a fair test, you want to be sure the only significantly changing things are the variables you intentionally change and the one that you are measuring in response to that change. Of course you cannot take into account everything, like air currents, but you should minimize those things as much as possible so they are not significant factor, in your results. Other possible variables you need to make sure are kept the same are the rate at which you pour the water onto the material or how you spray water on the material.

META NOTES

The student text asks students to present their procedures. Often students have to design their own procedures. In these investigations the procedures were provided for the students. Students will need to describe how they carried out the procedures consistently and if the experiment was a fair test. For example, in both investigations each group needs to determine a consistent way to pour or spray the water—same height and rate.

TEACHER TALK

"A fair test is when you compare things under the same conditions. If I was interested in how different items fall, I would want to drop those items from the same height and at the same time. It would not be a fair test if I dropped one of the objects from a higher or lower position. Similarly, I would want to be sure that I measured the time of flight in the same way. It would be really difficult for me to make sure there were not air currents or that they were the same at all times. I would need to make sure the doors and windows are closed and the air conditioning and/or heating is not on so I can minimize the effect of air currents. Although it might change, it is so small that it does not significantly affect my measurements. Of course, when I write up my results I'll have to mention all this."

○ Get Going

Tell students that each group will make a poster for the *Investigation Expo* and they should follow the bulleted list in the student text when they do this. Show students the sample poster you made, and how they can use the sample poster to organize all of the information that needs to go in their posters.

Point out the bulleted list of things about their investigation that their audience will listen for so they can keep these in mind as they prepare.

Distribute poster materials.

△ Guide

As groups are working on their posters and preparing for their presentations, assist them as needed. Some issues that may arise are listed below.

- Students might have trouble deciding how to determine confidence in their results. They can base this on the quality of their experiment by analyzing how consistently they ran their procedures, how they controlled variables, and how much variation was in their data.
- Students might not completely understand a fair test. You could describe it as only allowing two variables in their experiment to change: the variable they intentionally change (manipulated variable) and the one they measure in response to it (responding variable).
- Students might include opinions in their interpretations or claims. Ask if the claim is directly supported by their data or if something more is in their claim.
- Students might be using science knowledge incorrectly or incorrect information they think is accepted science knowledge.
- Remind groups everything they show on their poster or say in their presentation will have to be clear so that others can understand and follow their thinking.

META NOTES

There will be a class discussion after each group presents. The first two groups presenting should have the same experiment question so that you can compare the two experiment plans for the same experiment. After all four groups present you should compare experiment plans across experiments.

□ Assess

As you are visiting groups, decide which two groups from each investigation to present. You may want to pick groups that have very good examples or that have difficulties with fair tests or variables identification.

◇ Evaluate

Check to see if groups have completed the items required for their poster and ask if they are ready for presentations. Students should describe what makes or what would make their experiment a fair test on their posters. They should note if their measurements are good based on how clustered their trials are and the trend in their data. Their interpretation of results should agree with the trend in their data. Their claims should be supported by the data and contain no opinions and nothing that the data does not explicitly and directly support. If they are using science knowledge in their interpretations, it must be correct.

⬡ Get Going

Now that students have prepared their posters and presentations, they are ready to review each other's posters. Remind the class that for each poster they will need to look for what claims are being made and if they are believable. They can use the bulleted lists in the student text to help ask questions. While visiting posters, they should record questions they have for each group. They will have an opportunity to ask questions when groups present their experiment results. Each group should have both investigation pages to record data on from the posters they visit for future use, or you may want them to just record the trend from each.

Have each group display their posters around the room, and allow students to visit each poster for a minute or so to become familiar with each groups' work.

META NOTES

You should model what you expect from students. Encourage students to ask questions of the presenting group and the presenting group to respond to the student asking the question. Students should not make you the focal point during the discussion.

Questions should require more than one-word answers and should focus on a critical part of the investigation. Students will often want to question other students about color choice or layout. By modeling questions about content, you will assist students in asking better questions.

△ Guide a Class Discussion

Remind students that during the presentations they should be listening for descriptions of the results and claims. If they are unclear about the results or the claim they should ask questions of the presenting group. Emphasize that they should pay attention to consistent procedures and whether or not the experiment is a fair test.

Inform the two groups you selected for each experiment that they will be presenting. Remind students to provide their reasoning, describing why or why not their experiment is a fair test, and to address whether or not their results are in agreement or disagreement with the other results that have already been presented.

△ Guide Presentations and Discussions

Begin a short discussion after each presentation. Students should have an easier time leading the discussion since they have practiced discussions in the last section. Assist students with some of the language needed by modeling for them or asking questions that guide them:

- Do you agree with what ... said? Why? or Why not?
- I agree with .... because....
- I don't understand the reasoning behind your interpretation of your results. Could you start from showing the data and the trend?

Ask the class if they had questions about any of the posters involving this experiment question. Encourage discussion noting areas of difficulty students are having.

Point out that sometimes you cannot trust results because the procedures were not done it the same way.

After the first pair of groups presents (from the same investigation), ask questions to compare and contrast the investigation results and procedures. Since the students are following a set of procedures, the experiments should be fair tests provided they followed the procedures and controlled how they poured the water into the cup.

TEACHER TALK

"As I walked around while you were working, I noticed that group "x" was running their experiment like XYZ. I noticed that they made sure they ran their procedures the same way each time. How do you think?"

META NOTES

You can discuss the topic of trusting results —consistent procedures, fair tests, and variables— as soon as you have groups that have results that disagree. Otherwise, discuss this at the end of the discussion of the first investigative question.

META NOTES

Students often have a difficult time distinguishing between the types of variables. Often times, they think the manipulated (independent) variable is a control variable because they control its values. It is important to emphasize that control variables are held constant or kept the same.

NOTES

Be a Scientist: Different Kinds of Variables

Be a Scientist

Different Kinds of Variables

As you designed and ran your experiment, there were several kinds of variables you worked with:

- One that you changed or varied in your experiment. This is called the **independent variable** (or **manipulated variable**).
- Some were ones you worked hard to keep the same (constant) during every trial. These are called **control variables**.
- Some were ones you measured in response to changing the manipulated variable. These are called **dependent variables** (or **responding variables**). Their value is dependent on the value of the independent or manipulated variable.

Experiments are a very important part of science. When scientists design experiments, they think about the things that might have an effect on what could happen. They then identify exactly what they want to find out more about. They choose one factor as their independent (manipulated) variable. This is what they change to see what happens. They have to keep everything else in the procedure the same. The variables they keep the same, or hold constant, are control variables. Finally, there are factors they measure. These are the dependent (responding) variables. If they have designed a fair test, they can assume that changes in the dependent (responding) variables result from changes made to the independent (manipulated) variable.

If you ran Investigation 1, your independent (manipulated) variable was the type of material. If you ran Investigation 2, your independent (manipulated) variable was slope. In both experiments, your dependent (responding) variables were the distance the materials spread, the patterns of erosion, and the effects on the hill. Everything else, including the shape and size of the test container, amount of material tested, amount of water poured on each sample, and the way the water was poured, were control variables. To be sure that what was measured (the dependent or responding variable) was dependent on what was changed (the independent or manipulated variable), it was important to keep the controlled variables exactly the same every time a trial was run.

independent (manipulated) variable: in an experiment, the variable the scientist intentionally changes.

control variables: in an experiment, the variables kept constant (not changed).

dependent (responding) variables: in an experiment, the variables whose values are measured; scientists measure how these variables respond to changes they make in a manipulated variable.

△ Guide

Using one of the presented experiments, point out the manipulated variable (what you intentionally vary — materials or slope) and the responding variable (how far the material is displaced). Students who ran *Investigation 1* may not recognize that they varied the materials. Because they used the same materials in all three trials. They separately examined how far particles of each material were displaced in each trial, so they varied the materials within each trial.

Point out all of the variables that could have been changed but were not (amount of water poured, height of hills, and amount of material used). These are called the control variables.

Discuss the claims made by the groups and how confident they are. Students should be confident if their claims are based on the trends of their data and if their data results from a fair test in which procedures were followed consistently. Compare the claims of both groups. Although groups for each investigation will not have identical data, they should see similar trends in their data resulting in similar claims.

TEACHER TALK

"Let's consider the claims. How do we determine if the claim is good? Do you trust the claims the groups made? Do you think we can make a claim based on both experiments?"

Point out that scientists call claims valid if many different groups see similar trends when they are investigating the same variables (e.g., similar displacement for materials and similar slopes).

NOTES

Reflect

15 min.

Students participate in a class discussion and then reflect on their own investigations, fair tests, and believable claims and how their claims will help solve the Basketball-Court Challenge.

Reflect

Answer the following questions. This will prepare you for a class discussion about what you now know that will help you achieve the *Basketball-Court Challenge*. Be prepared to discuss your answers with your class.

1. What variable were you investigating in your experiment? What were you investigating about that variable? How did you vary it to determine its effects?
2. List all of the variables you tried to hold constant in your experiment.
3. How many trials did you perform? Why did you perform that number of trials? Was this a good number of trials?

4. For those who did *Investigation 1*: How consistent was your set of data? Why is consistency in repeated trials important in an experiment?

   For those who did *Investigation 2*: How consistent was your data with the data of other groups who ran the same investigation? Why is it important for your data to be consistent with the data collected by other groups?
5. How useful was your data in determining the affect of your variable on erosion?
6. What do you think you now know about the effects of particle size on erosion that will help you design a way to control erosion at the basketball court? What do you think you know about the effects of slope on erosion that will help you control erosion at the basketball court?

⬡ Get Going

Ask groups to come up with their best group response to the six questions.

Let groups know how much time they have for answering the questions. You should give no more than 10 minutes.

TEACHER TALK

"In our *Investigation Expo,* we reviewed everyone's results on posters, listened to four presentations, and discussed some important ideas like consistent results and fair tests. Now it's time to think about how all you know now pertains to your own investigations and to the challenge."

☐ Assess and Guide

Hold a discussion using the information provided below and guide students using questions that lead them to the essential part of the answer. Answers to some or all of Questions 1-4 might have already come up during the discussion. Use these questions and responses to lead students to think about Questions 5 and 6. Spend most of the discussion time focused on why their data is useful (Question 5) and how they might be able to design a better way to control erosion around the proposed basketball court (Question 6).

Assess if students understand what they were investigating, the variables involved, and what made their results trustworthy. Use these to assist students in recognizing how their experiments will help solve the challenge of recommending an erosion-control design for the proposed basketball court. Student responses should contain the following:

1. In *Investigation 1,* students were investigating the material type variable. They examined how much each type of material was displaced separately. In *Investigation 2,* students were investigating the slope variable. They changed the elevation of the hill to vary this.
2. For both investigations, students should have used the same amount of water and poured or sprayed from the same distance. They should have used the same amount of test materials and made their hill the same way each time.
3. Students performed three trials. Students will probably say they had no choice in the number of trials; they did what they were told. Ask students why they think they were asked to do three trials. They should note that the more trials they do, the better they can determine how precise their measurements are.
4. For those who ran *Investigation 1,* without consistency in repeated trials, it is not possible to determine what the results mean and whether they are reliable. If a lot of trials are taken and the results

are not consistent, it indicates a problem with the procedures or measuring tools. For those who ran *Investigation 2,* consistency between different groups running the same investigation is necessary to determine whether the results are reliable. If all of the groups get different results, there is a problem with the procedures or measuring tools.

5. Students should recognize that the usefulness of the results depends on how reliable the results were. Their responses may also discuss how well their investigation modeled erosion in natural conditions.

6. This question is meant to help students connect the results of these investigations to the criteria of the challenge. Students should recognize the utility of knowing how different materials erode. They may be able to support the area around the basketball court with a material that is resistant to erosion or pick an erosion control method that works best with the material of the hillside. They should discuss the claims that groups made and should connect these to the challenge.

META NOTES

Consider using the class *Project Board* as a way to summarize what students have learned in this section. Have students fill out what they have learned (claims) and their evidence (observations from their experiments) in Columns 3 and 4.

NOTES

Slope, Particle Size, and Erosion

Erosion moves soil and other particles. Force is needed to move anything; the main driving force of erosion is gravity. Gravity can move sediments by acting on them directly. Pieces of rock on cliffs and steep slopes, broken loose by weathering, fall or slide downhill under the direct influence of gravity. Gravity can also move sediments by acting on them through agents of erosion. If water runs downhill under the direct influence of gravity, the running water can then exert an indirect force on rock particles in its path, causing them to move. The running water is an agent of erosion. Other agents of erosion include winds, glaciers, waves, and ocean currents.

The faster water moves, the more force it can exert. With more force, water can move more rock particles and larger ones. The speed at which water flows downhill is directly affected by the slope, or steepness, of the land. The steeper the slope, the faster the water flows downhill, and the greater its power of erosion.

Water moving at different speeds can move different-sized particles. If water moves at 50 cm/s, it exerts enough force to move sand particles (and anything smaller) but not pebbles. At that speed, with water flowing over a mixture of sand and pebbles, the sand will be carried downstream, but the pebbles will be left behind. This way, running water can cause a mixture of different-sized particles to become sorted, or separated, according to their size.

What's the Point?

Your class completed two investigations to answer two different questions. One half of the class collected data about how particle size affects erosion of soil and other materials. The other half of the class collected data about the effects of slope on the erosion of soil and other materials. Each group then interpreted their results by identifying trends in the data and stating a claim based on those trends. When everyone was finished, each group shared what they found in an *Investigation Expo*. By sharing results, everyone was able to get the information needed to answer both questions. This is the way scientists work. Presenting results of investigations to other scientists is one of the most important things they do. This lets other scientists build on what they learned. You interpreted the data from your investigation. The trends you found and the claim you made will help you in achieving the *Basketball-Court Challenge*.

Teacher Reflection Questions

- What difficulties did students have identifying the variables in their investigations? What can you do to help them with this in future investigations?
- How were you able to help students find trends in their data? How could you help students with this in the future?
- How were you able to keep students focused on their observations as they ran their investigations? Is there anything you would do differently next time?

NOTES

SECTION 3.4 INTRODUCTION

3.4 Explain

Create an Explanation

◀ ***1 class period***

A class period is considered to be one 40 to 50 minute class.

Overview

Students are introduced to scientific explanations and how to create them. They practice creating explanations for erosion based on their observations and science knowledge from the last section. Students share their explanations with the class. By hearing others' explanations students will be helped to refine their own in the next section.

Targeted Concepts, Skills, and Nature of Science	Performance Expectations
Scientists often work together and then share their findings. Sharing findings makes new information available and helps scientists refine their ideas and build on others' ideas. When another person's or group's idea is used, credit needs to be given.	Students should be able to describe how their ideas are changing.
Scientists make claims (conclusions) based on evidence obtained (trends in data) from reliable investigations.	Students should be able to make claims based on trends in their data and in the class's data. This could be part of their explanation.
Explanations are claims supported by evidence, accepted ideas, and facts.	Students should be able to create an explanation of why erosion happens the way it does based on the results of all the class experiments and what they learned in the previous section.
Erosion is the process of soil and other particles being displaced by water, waves, wind, and gravity.	Students' claims and explanations should reflect their understanding that erosion is the process of Earth materials being displaced by water.

Materials	
1 per student	*Create Your Explanation* page

Homework Options

Reflection

- **Science Process:** Identify the claim, evidence, and science knowledge in the following explanation:

  When the natural vegetation in an area is destroyed, it can lead to accelerated erosion. This is because the vegetation often holds the topsoil in place and slows erosion. This was observed during the Dust Bowl in Case Study 4. *(Claim: When the natural vegetation in an area is destroyed, it can lead to accelerated erosion. Evidence: This was observed during the Dust Bowl. Science knowledge: This is because the vegetation often holds the topsoil in place and slows erosion.)*
- **Science Content:** Using data from the class's investigations, explain why the landslide at Laguna Beach happened on a steep slope. *(Students can use the results of* Investigation 2. *Their data should show that the steeper a slope is, the more likely it is that there will be a landslide when the ground is soaked.)*

Preparation for 3.5

- **Science Content:** Based on what you now know, propose a method to control erosion around the basketball court. You should consider what kinds of materials you might use, how the slope around the court affects erosion, and how rainfall affects erosion. *(This question is meant to get students applying their knowledge of erosion and thinking about erosion control in the context of the challenge. It also prepares them for the remainder of the Unit, which focuses on erosion control.)*

SECTION 3.4 IMPLEMENTATION

◀ *1 class period**

3.4 Explain

Create an Explanation

After scientists get results from an investigation, they try to make a claim. They base their claim on what their evidence shows. They also use what they already know to make their claim. They explain why their claim is valid. The purpose of a science explanation is to help others understand:

- what was learned from a set of investigations or case studies, and
- why the scientist reached this conclusion.

Later, other scientists use these explanations to help them explain other phenomena. Explanations can also help them predict what will happen in other situations.

You will do the same thing now. Your claims and explanations will be about what causes erosion. Each group will use the examples found on your *Erosion Walk*, information from the case studies you have read, and evidence from the experiments you just completed to make a claim about why erosion happens. You will then create an explanation to support your claim. You will be reporting your explanation to your classmates. With a good explanation that matches your claim, you can convince them that your claim is valid.

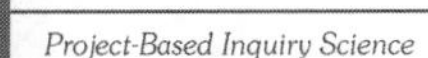

Because your understanding of the science of erosion is not complete, you may not be able to fully explain the causes of the erosion you observed. But you will use the evidence you have collected and what you have read to come up with your best explanation. Scientists finding out about new things do the same thing. When they only partly understand something, it is impossible for them to form a "perfect" explanation. They do the best they can based on what they understand. As they learn more, they make their explanations better. This is what you will do now and what you will be doing throughout PBIS. You will explain your results the best you can based on what you know. Then, after you learn more, you will make your explanations better.

3.4 Explain

Create an Explanation

15 min.

Students are introduce to scientific explanations which they will be constructing throughout the PBIS *curriculum.*

META NOTES

Many students will already have ideas about what an explanation is. In science and *PBIS*, an explanation is a claim that is supported by evidence and science knowledge in a logical way.

○ Engage

Briefly review some of the observations students made in their investigations and remind them of what they learned from their case studies. Tell students that when they make recommendations to the school, they will need to explain how factors in the environment will affect erosion around the basketball court. The school will want a scientific explanation.

*A class period is considered to be one 40 to 50 minute class.

Be a Scientist: What Do Explanations Look Like?

10 min.

Be a Scientist

What Do Explanations Look Like?

Making claims and providing explanations are important parts of what scientists do. An explanation connects three parts:

- Claim—a statement of what you understand or a conclusion that you have reached from an investigation or set of investigations.
- Evidence—data collected during investigations and trends in that data.
- Science knowledge—knowledge about how things work. You may have learned this through reading, talking to an expert, discussion, or other experiences. Science knowledge comes from investigations others have done over many years. It is knowledge that scientists agree about.

An explanation is a statement that connects the claim to the evidence and science knowledge in a logical way. A good explanation tells what causes the statement in a claim. The best scientific explanations use agreed-upon science knowledge in a logical way to support a claim. These kinds of explanations can usually convince others that a claim is valid.

For example, suppose you live in a city in the USA that gets cold and has snow in the winter. It is fall. You see a lot of birds flying past your home. You wonder why so many birds are flying by. You have learned that many birds cannot live in cold places. They fly to warm places (usually south) to spend the winter. You wonder if these birds are flying by your home on their way to a warmer place. You take out your compass and observe that the direction they are flying is south. You conclude that the birds are flying past your home to a warmer place where they will spend the winter. Look at how you can form an explanation.

Your claim: The birds flying past my house are flying south for the winter.

Your evidence: The birds are flying in a southern direction. (You have observed and measured that using a compass.) It is autumn.

Your science knowledge: Many birds cannot live in cold weather. Birds that cannot live in cold weather fly to warmer climates when the weather begins to cool. They stay there for the winter.

TEACHER TALK

"When you make your recommendations to the school, you will need to explain factors in the environment that affect erosion around the basketball court. From your investigations, you have evidence about how different factors affect erosion. You also have a lot of information about erosion from the case studies. You will need to use this evidence and science knowledge to explain your claims about erosion. That way, when you make your recommendations to the school, you will be able to explain your reasons."

△ Guide

Tell students that a scientific explanation is a statement that connects a claim to the evidence and science knowledge supporting the claim. The connection has to be made in a logical way.

TEACHER TALK

"When scientists make an explanation, they are answering a question about a situation or phenomenon. They use evidence and science knowledge to make a claim. They also use logic to connect their evidence and science knowledge to their claim. An explanation is not based on opinions. It is O.K. to not have complete or perfect explanations if you do not have complete or perfect understanding. Just like scientists, you should explain things as well as you can, based only on what you know.

In the *Lava-Flow Challenge* you had a claim that you could support with evidence. If the volcano erupts, the lava could flow with a speed of 3 meters per second. This is supported by my data, which showed that when I poured model lava down a model slope, on the average it flowed at a rate of 3 m/s. I do not have enough information yet to explain what causes it to flow at this rate.

This explanation does not state why the lava flows at the rate it does. If you had more information, you could explain why."

META NOTES

The process of creating an explanation is complex. Students will need scaffolding with plenty of examples as they begin learning how to construct explanations. In *PBIS*, explanations are usually revisited and revised a number of times as more information is obtained. After creating their explanations, students will share their explanations with the class and have an opportunity to revise them based on feedback from the class.

Review what claims, evidence, and science knowledge are. Emphasize that a claim is a statement or conclusion reached from one or more investigation. Evidence is the data from the investigation. Science knowledge is knowledge about how things work based on previous investigations by experts. Use the example of the lava flow to illustrate each component. Point out that the lava example showed the claim is the conclusion of the experiment and the evidence is based on the trends in the data.

TEACHER TALK

"Let's consider the claim and evidence from the lava example.

Claim: The lava flows at a rate of 3 m/s.

Evidence: We performed experiments and, on the average, measured 3 m/s for the flow of the model lava over the model land.

Scientific Knowledge: I don't have enough information to explain why lava flows at the rate it does.

Here the claim is basically just the conclusion to the experiment because there is no scientific knowledge available to make a bigger claim, so you can only make claims on what you have actually measured or observed."

Use the example explanation about birds flying south in the student text to point out how each component (claim, evidence, and science knowledge) is connected in a logical way. When students say their explanation they may just say the first line, but when they write their explanation they should connect all the parts together to support the claim.

Below is an additional example illustrating an explanation of falling. You may want to demonstrate the items being dropped in the evidence.

Claim: All things with mass fall to ground when they are dropped.

Evidence: Everything I have seen dropped—a ball, a piece of paper, a pen, an eraser.

Science Knowledge: I have read about the force due to gravity. Mass attracts mass and this attraction increases as the mass increases. The effect of this force is not easily observed unless one of the objects is very massive, like Earth.

Explanation: Dropped objects fall because the force due to gravity between the object and Earth pulls the object towards Earth. Everything I have seen dropped in class (a ball, a piece of paper, a pen, an eraser) support this.

NOTES

Your explanation: The birds flying past my house are flying south for the winter. Many birds cannot live in cold weather. Birds that cannot live in cold weather fly to warmer climates when the weather begins to cool. These birds are probably birds that need to go to warmer places in the winter. They are flying south to find a warmer place to stay while it is winter here.

An explanation is what makes a claim different from an opinion. When you create an explanation, you use evidence and science knowledge to back up your claim. Then people know your claim is not simply something you think. It is something you have spent time investigating. You have found out things that show your claim is likely to be correct.

Explain

Writing an Explanation

Here is an example from the *Boat Challenge* that might help you understand more about writing an explanation. When you worked on that challenge, you saw that boats designed to push down on a large area of water and to hold air in their structures were able to keep afloat even with the weight of the keys. You read that a flat piece of foil has more buoyant force than a foil ball because a greater area of water is pushing up on it. You also read that having air as part of the structure of the boat decreases the overall density of the boat and its cargo and increases the buoyant force of the boat. Look at the claim and the explanation that could have been created by a group of students who worked on the *Boat Challenge*.

Some boats, such as this barge, are designed to carry huge amounts of cargo and still float easily through the water. If the boat is less dense than the water it displaces, the boat will float.

Explain

up to 15 min.

Students are introduced to the process of constructing an explanation and then construct their own for erosion.

△ Guide

Remind students of the evidence and science knowledge of surface area and density of boats from the boat challenge. Note that the explanation in the student text is a recommendation. A recommendation is a type of claim based on evidence and science knowledge. Students will learn more about recommendations in *Section 3.8*.

Walk through the example of constructing an explanation, describing the reasoning behind each step. Emphasize that there are no opinions in explanations.

Look at the claim. Remember that a claim is a statement of proposed fact that comes from an investigation.

Claim – a statement of what you understand or a conclusion that you have reached from an investigation or a set of investigations.

A boat with a larger surface area and a hollow compartment to hold air will float better than a boat with a smaller surface area and no compartment to hold air.

This claim comes from looking at the trend they found in the data they collected.

Evidence – data collected during investigations and trends in those data.

Flat foil boats with hollow compartments were able to hold 8 keys and to float for 20 seconds. Boats that were not flat or did not have hollow compartments did not hold as many keys and sometimes sank.

They then read about the science of buoyant force and summarized the science knowledge they gained.

Science knowledge – knowledge about how things work. You may have learned this through reading, talking to an expert, discussion, or other experiences.

As the surface area of a boat increases (when its weight stays the same), it pushes down on a larger area of water. When this happens, the boat is in contact with more water molecules that push up on the boat, and the water can better support the boat. If the force pushing up from the water is greater than or equal to the force pulling down on the boat by Earth, the boat will stay afloat.

When you place a boat in water, it pushes away some water. When the overall density of the boat and its cargo are less dense than the water that displaces it, the boat will float.

Air has a much lower density (mass per volume) than water. If air is included in the overall materials making up the boat, that air could decrease the overall density of the boat and help make it more likely to float.

⬡ Get Going

Have students work in groups to construct their own explanations of what causes erosion using their claims, data, and the science knowledge presented in this Unit. The science knowledge should be drawn from the case studies. (The Laguna Beach study will be especially useful.) Students should use their *Create Your Explanation* pages.

Let groups know that they will be presenting their explanations to the class.

Here is a logical statement that ties the claim, evidence, and science knowledge together. Notice how the claim, the evidence, and the science knowledge are all part of this explanation.

> ***To hold a lot of mass, we recommend designing a boat that covers as much surface area as possible and that has places in its structure that hold air. Our data showed that the boats with larger surface areas and the ability to hold air remained afloat even with the increased weight of the cargo. We know that, to increase the buoyant force of an object, you can spread the mass of the object over a greater area. The larger surface area is in contact with a greater number of water molecules that exert an upward push on the boat, providing more support. We also know that air has a very low density and can decrease the overall density of an object. You can decrease the downward push of the boat by decreasing its density. As long as the push down from the boat is not greater than the push up from the water, the boat will float, even with cargo.***

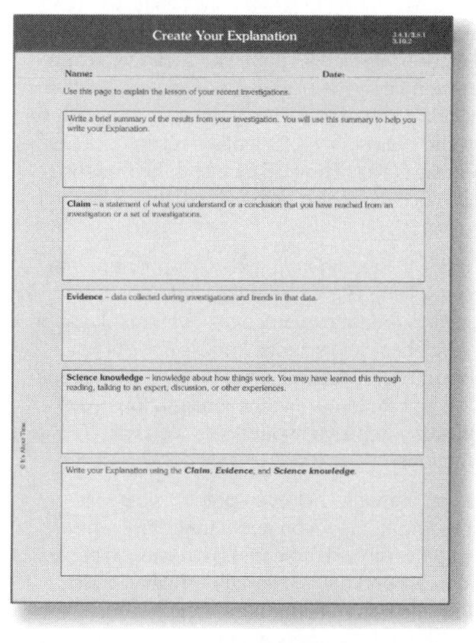

Create Your Explanation

Name: ______________________ Date: __________

Use this page to explain the lesson of your recent investigations.

Write a brief summary of the results from your investigation. You will use this summary to help you write your Explanation.

Claim – a statement of what you understand or a conclusion that you have reached from an investigation or a set of investigations.

Evidence – data collected during investigations and trends in that data.

Science knowledge – knowledge about how things work. You may have learned this through reading, talking to an expert, discussion, or other experiences.

Write your Explanation using the ***Claim***, ***Evidence***, and ***Science knowledge***.

This may seem like a long explanation. Long explanations are not always needed. However, seeing this explanation may help you as you try to write an explanation. You will use a *Create Your Explanation* page, similar to the one shown, to help you with explanations. It will give you space to write your claim, your evidence, and your science knowledge. It will also remind you what each of these is.

☐ Assess

As groups are constructing their explanations, assess how they are doing paying particular attention to see if they have included opinions and if they are making logical connections (they should use phrases like, “this follows because...”).

△ Guide

While groups are working on their explanations, assist individual students and groups as needed. If you notice the majority of the class having difficulty in constructing a particular part of an explanation, then stop the class and hold a class discussion.

Communicate

up to 15 min.

Students hold a class discussion on explanations presented by groups.

META NOTES

Sharing explanations helps students to articulate. Hearing their classmates' explanations will help them to refine their own. This is a perfect time to encourage discussion of various ideas. The process of thinking about the explanations and weeding out solid evidence from non-specific evidence, is part of the process of doing science. Sometimes, students have difficulty when they are asked to engage in these discussions. Encourage students to discuss rather than fight for their positions. Encourage them to provide evidence for their explanations and to consider the evidence of others.

Consider creating a class list of explanations to refer back to later.

Communicate

Share Your Explanation

When everyone is finished, you will share your explanations with the class. As each group shares theirs, record the explanation. You might also create a poster for the classroom that has the full set of explanations on it. You will have an opportunity to revise your explanations after you learn more about what causes erosion and how it can be managed.

What's the Point?

Science is about understanding the world around you. Scientists gain understanding by investigating and explaining. The results of investigations are useful in making sense of and organizing the world. To help others better understand what they have learned through their investigations, scientists must communicate their results and understandings effectively. Scientists make claims about the phenomena they investigate. They support their claims with evidence they gather during investigations. They also read science that others have written about. They combine all of that together to create explanations of their claims—statements about why their claims are so. Other scientists carefully examine these explanations. They discuss them with each other. They try to decide if the explanation is complete enough for them to be sure about whether the claim is valid. Scientists accept a claim as valid when many different scientists agree. The evidence and their science knowledge must justify the claim. Scientists also help each other make their claims and explanations better.

Throughout this school year, you will investigate a variety of phenomena. You will apply what you learn to solving *Big Challenges* and answering *Big Questions*. You will be asked to create explanations. Every explanation you write will include a claim, evidence, and science knowledge. As you move through each Unit and learn more, you will create new explanations. You will have the opportunity to edit and improve the explanations you created earlier. Just as you iteratively improved your boats, you will iteratively improve your explanations.

You will also use explanations you create to help you predict what will happen in new situations. For example, now you know that rain, waves, gravity, and wind are all causes of erosion. That means you can probably predict what might cause erosion around the proposed basketball court. Make sure you can make that prediction. Making that prediction successfully will help you know that you understand the science you have been learning.

⬡ Get Going

Let the class know that if they are in the audience, they should be looking for the parts of an explanation. Model for them how to ask for clarification.

TEACHER TALK

"I don't understand how the science knowledge backs up your claim. Could you walk me through it?"

Let the class know which set of groups (pertaining to a particular investigation question) will be presenting first and then begin the presentations.

△ Guide

After each group presents, hold a class discussion on their explanation. You may want to have the class pick out what the claim, the evidence, and the science knowledge if the presenting group is not getting many questions.

Encourage the presenting group to answer the person asking the question.

At the end of the presentations and discussions, let students know that they will have a chance to revise their explanations during the next section. Then summarize the parts of an explanation.

Below are example explanations that you can use as a reference for yourself and to help guide students. Some of the science content is not in the reading so students may have difficulty incorporating it in their explanations. This is O.K. because the focus of this Unit is the practices of *PBIS* and the content will be visited again in a later Unit.

META NOTES

You may only want to have two groups from each investigative question present. Pick two groups that have not presented before.

Example Explanations

Explanation 1: When water runs over land, Earth materials composed of smaller particles are eroded more than Earth materials composed of larger materials. We know this because we ran water over a mixture of different Earth materials. The Earth materials composed of small particles were moved more than the Earth materials composed of large particles.

The claim is when water runs over land, Earth materials composed of smaller particles are eroded more than Earth materials composed of large materials. The evidence is that students observed water move small particles more than large particles in *Investigation 1.*

Explanation 2: When land is saturated, steep slopes are eroded more than gentle slopes. We know this because we saturated slopes of varying steepness and observed that the steeper the slope, the greater the erosion.

The claim is that saturated land erodes more at steep slopes than at gentle slopes. The evidence is that students observed steep slopes erode more than gentle slopes.

☐ Assess

Assess students' ability to construct and pick out claims, evidence, and science knowledge and their ability to construct explanations. Remember that they are not to master this yet.

◇ Evaluate

Students should have explanations concerning how water passes through materials and over materials and how it displaces those materials.

Students will be writing explanations throughout *PBIS*. Writing good explanations is a key part of doing science. It is important for them to understand the parts of an explanation and to be able to construct one. Writing explanations is not an easy task. They will have a chance to revise their explanations during the next section.

Highlight the term valid. Scientists say a claim or a test is valid if there is repeatable evidence supporting it, no evidence against it, and if it is logical.

Teacher Reflection Questions

- What difficulties did students have understanding what a scientific explanation is? What ideas do you have for guiding their understanding during the next section?
- Would modeling another explanation be beneficial for the students? What ideas do you have for modeling explanations?
- How did you use the student text? How could you assist students in their reading ability as well as their comprehension?

NOTES

NOTES

SECTION 3.5 INTRODUCTION

3.5 Explore

1 class period ▶

A class period is considered to be one 40 to 50 minute class.

What Are Some Ways Erosion Can Be Managed?

Overview

Students explore solutions to erosion problems, analyzing and evaluating case studies of successful erosion control. The class discusses a case study together, and then each group independently analyzes one of four case studies. They present the case studies to the class and discuss what made each solution effective. Groups then reflect on how each erosion-control method is similar to things they have seen firsthand and how it can be applied to the *Basketball-Court Challenge*. Finally, the class updates the *Project Board* with what students think they know about controlling erosion and what they need to find out.

Targeted Concepts, Skills, and Nature of Science	Performance Expectations
Scientists often work together and then share their findings. Sharing findings makes new information available and helps scientists refine their ideas and build on others' ideas. When another person's or group's idea is used, credit needs to be given.	Students should work in small groups analyzing case studies of erosion control. Through sharing their results with the class, students should build on each other's results and ideas.
Identifying factors that could affect the results of an investigation is an important part of planning scientific research.	Students should identify the variables that contributed to erosion in the cases they read.
Erosion is the process of soil and other particles being displaced by water, waves, wind, and gravity.	Students should be able to describe ways to control erosion after analyzing cases of erosion and erosion control.

Materials	
5 per student	*Case Summary* page
1 per class	Class *Project Board*

Homework Options

Reflection

- **Science Content:** How are the situations you read about and discussed in these studies similar to the situation you face in the *Basketball-Court Challenge?* How are they different? *(Students should recognize all of the case studies involved streams of running water—tributaries and creeks. In the* Basketball-Court Challenge *there is no permanent stream. They all involve Earth materials eroding from a slope due to the effects of water and gravity. Similarly, the erosion around the basketball court is from a slope and will be exacerbated by water and gravity.)*
- **Science Content:** What questions about erosion occurred to you as you read or discussed the case studies? *(This question is intended to engage students in thinking about what they have yet to learn about erosion.)*

Preparation for 3.6

- **Science Content:** Which of the erosion-control methods that you discussed today do you think would be effective in the *Basketball-Court Challenge?* Why? *(Students should support their claims with reasons.)*

1 class period ▸

3.5 Explore

What Are Some Ways Erosion Can Be Managed?

5 min.

Students learn more about erosion control in real-life situations by examining case studies.

SECTION 3.5 IMPLEMENTATION

3.5 Case Studies

What Are Some Ways Erosion Can Be Managed?

You are probably aware that the problem surrounding your proposed basketball court has to do with eroding soil material due to water and gravity. You are now ready to start looking for ways to control the erosion. Knowing about erosion-control methods that have been used by others may help you solve your problem at the basketball court.

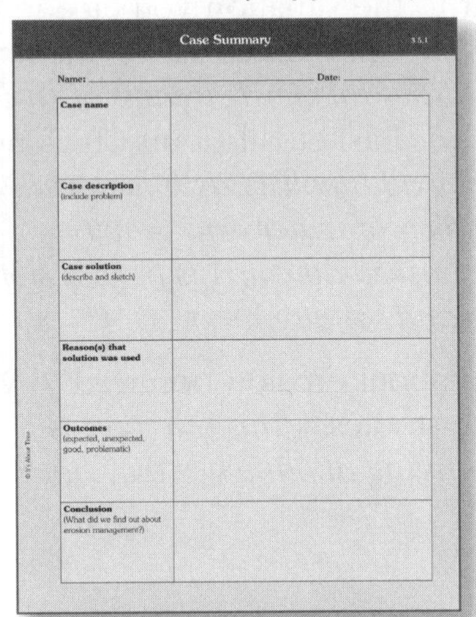

Case Summary 3.5.1

Name: ______ Date: ______

Case name	
Case description (include problem)	
Case solution (describe and sketch)	
Reason(s) that solution was used	
Outcomes (expected, unexpected, good, problematic)	
Conclusion (What did we find out about erosion management?)	

You will read about different erosion-control methods and see what you can find out. You may also have some ideas about how to control erosion based on what you have seen in your own neighborhood or around your community. Real-life examples that you are familiar with can be cases, too. You will have the opportunity to share these examples after everyone has read the cases that are coming up, and your class is ready to update the *Project Board*.

Coastal erosion is a natural process that results from precipitation, wind, and the constant movement of water, sand, and rock. Because communities have designed buildings very close to beaches and on cliffs above oceans, coastal erosion has damaged many structures and has put many buildings in danger of collapse.

◯ Engage

Begin by asking students what successful models of erosion control might look like. Where would they go to see examples of erosion control? What would they look for in those examples that they could apply to the *Basketball-Court Challenge?*

On the next few pages, you will find pictures and text about erosion control. You will soon have the chance to model some of these techniques. Read all of the case studies, and work with your group to complete a *Case Summary* page as you review each case. Some of these cases deal with problems that are similar to the one you are facing at the basketball court. You should look for these similarities and pay attention to how the problem was solved.

Case Study 1: Boggy Creek Tributary

The storm discharge from Boggy Creek Tributary at the Poguito Street culvert (a drain passing under a road) had washed away a portion of the land on either side of the channel. The land was quickly eroding away, and the channel was getting very close to a nearby house. This project rebuilt and strengthened the bank, or side, of the channel, using limestone blocks to build a wall. The wall prevented the water from further eroding the soil. The yard was restored and the home is now protected.

You can see the damage due to erosion before the problem at the Poguito Street culvert was corrected.

Building a wall made of limestone blocks restored the area and prevented any further erosion.

TEACHER TALK

"From your experiments, our erosion walk, and the case studies you read about, you probably have a good idea what erosion can look like. What do you think might be good solutions to erosion? What if you wanted to find out? Where would you go? Are there any situations like the *Basketball-Court Challenge* where you could see solutions that would help you? Think about what you would look for in those solutions."

△ Guide

Let students know they will read case studies about successful solutions to erosion. Emphasize that students should pay attention to what did not work well, what did, and to the conditions that are similar and different from the *Basketball-Court Challenge*.

Draw students' attention to *Case Study 1: Boggy Creek Tributary*. Ask them what was happening in the land depicted in the top picture, and what has been done to the land in the bottom picture. Read the text with the class and discuss why this erosion control might be effective. Model for students how to go through the case study by analyzing the first case study as a class.

TEACHER TALK

"Looking at this case study, why do you think the wall of limestone bricks worked? Is it a good solution? Why? What were the reasons for building the wall?"

Distribute *Case Summary* pages and guide students through completing one for this study. Elicit the students' responses and ask students how the response answers the question on the *Case Summary* page. Students' answers should be similar to following examples.

NOTES

Case Summary

3.5.1

Name: ______________________ Date: ____________

Case name	*Case Study 1: Boggy Creek Tributary*
Case description (include problem)	*Storm discharge was washing away land on both sides of a bank, very close to a nearby home.*
Case solution (describe and sketch)	*The solution was to rebuild and strengthen the channel bank using limestone to build a wall.* *Students' sketches will vary but they should clearly show the limestone wall supporting the channel bank.*
Reason(s) that solution was used	*The wall was stronger than the dirt banks and able to protect the soil banks from eroding further.*
Outcomes (expected, unexpected, good, problematic)	*The wall prevented the water from further eroding the soil.*
Conclusion (What did we find out about erosion management?)	*Walls can help prevent erosion caused by water acting on soil banks.*

⬡ Get Going

Let the class know if they are in the audience (not presenting) they should be looking for the parts of an explanation (claims, experimental evidence, and science knowledge with no opinions) and model for them how to ask for clarification.

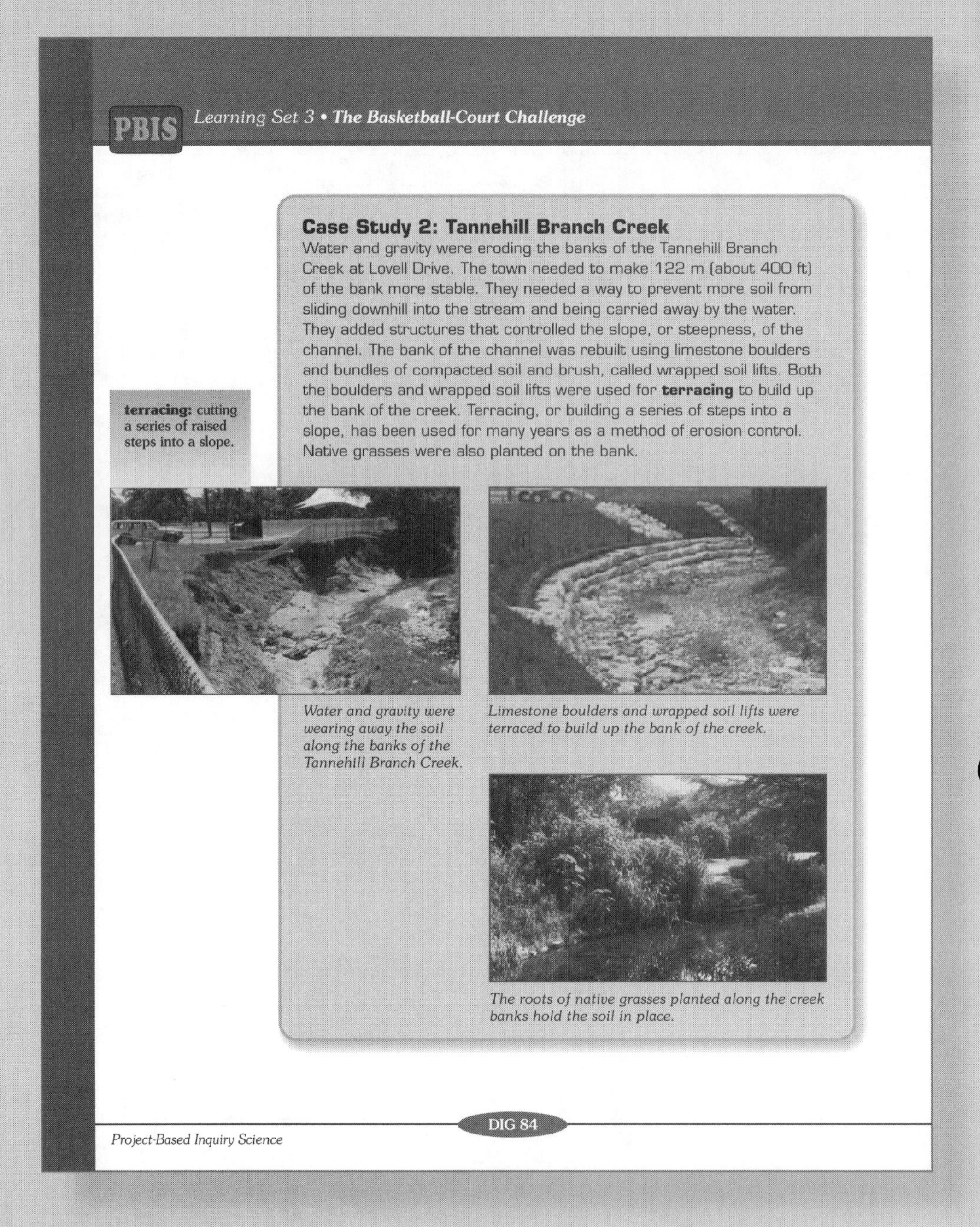
PBIS Learning Set 3 • The Basketball-Court Challenge

Case Study 2: Tannehill Branch Creek

Water and gravity were eroding the banks of the Tannehill Branch Creek at Lovell Drive. The town needed to make 122 m (about 400 ft) of the bank more stable. They needed a way to prevent more soil from sliding downhill into the stream and being carried away by the water. They added structures that controlled the slope, or steepness, of the channel. The bank of the channel was rebuilt using limestone boulders and bundles of compacted soil and brush, called wrapped soil lifts. Both the boulders and wrapped soil lifts were used for **terracing** to build up the bank of the creek. Terracing, or building a series of steps into a slope, has been used for many years as a method of erosion control. Native grasses were also planted on the bank.

terracing: cutting a series of raised steps into a slope.

Water and gravity were wearing away the soil along the banks of the Tannehill Branch Creek.

Limestone boulders and wrapped soil lifts were terraced to build up the bank of the creek.

The roots of native grasses planted along the creek banks hold the soil in place.

Project-Based Inquiry Science DIG 84

⬡ Get Going

Assign each group one of the four remaining case studies. There should be two groups analyzing each case study. Give groups about ten minutes to work on these case studies.

☐ Assess

As groups work on their case studies, ask them how the erosion problem in their case was solved. Does the solution give them any ideas for the *Basketball-Court Challenge?*

Case Summary

3.5.1

Name: ______________________ Date: ____________

Case name	*Case Study 2: Tannehill Branch Creek*
Case description (include problem)	*Bank erosion caused by water and gravity*
Case solution (describe and sketch)	*The town added structures that controlled the slope of the channel. The bank was rebuilt with limestone boulders and wrapped soil lifts. Native grasses were planted to help reduce erosion.* *Students' sketches will vary but should clearly show the limestone boulders and wrapped soil lifts supporting the bank and the vegetation planted to hold the soil.*
Reason(s) that solution was used	*The wall and wrapped soil lifts helped to strengthen the banks and change the slope of the bank. These should reduce erosion. Planting vegetation also reduces erosion.*
Outcomes (expected, unexpected, good, problematic)	*The wall stabilized the bank. Roots of the vegetation helped to hold the soil in place.*
Conclusion (What did we find out about erosion management?)	*Walls and vegetation can help prevent erosion caused by water and gravity acting on soil banks.*

NOTES

Case Summary

3.5.1

Name: ______________________ Date: ____________

Case name	*Case Study 3: Shoal Creek*
Case description (include problem)	*Bank erosion by high waters and obstructions in the stream channel caused a fallen tree. An oak tree and a house were threatened by erosion occurring.*
Case solution (describe and sketch)	*Rebuilding the creek bank with concrete, wrapped soil lifts, geogrids, and native grasses in a terraced fashion stabilized the bank.* *Students' sketches will vary but should clearly show the wall of concrete, wrapped soil lifts and geogrids. Vegetation should also be included to show the soil being held together.*
Reason(s) that solution was used	*The wall was stronger than the dirt banks and able to protect the soil banks from eroding further.*
Outcomes (expected, unexpected, good, problematic)	*The wall and terracing stabilized the bank and prevented further erosion.*
Conclusion (What did we find out about erosion management?)	*Walls and terracing can help prevent erosion caused by water acting on soil banks.*

NOTES

Case Study 3: Shoal Creek

There were several problems along the Shoal Creek bank. High waters from storm flows, obstructions in the stream channel, and erosion along the hillside banks contributed to the erosion problem near Pembrook Drive. A large oak tree had slid down into the creek and rested on the bottom of the channel. Another live oak, as well as a house, was threatened by further erosion. In this situation, the creek bank was rebuilt with concrete, wrapped soil lifts, and native grasses. After the project was completed, the bank was stable, and the natural stream setting was attractive. A house and oak tree were also protected.

Shoal Creek before the channel was cleared and the banks made stable.

Erosion-control methods used at Shoal Creek made the stream banks stable, preventing further land loss.

NOTES

Case Study 4: Little Walnut Creek

In the early 1980s, a property owner constructed a stone wall to hold back the soil and prevent it from sliding downhill into Little Walnut Creek. In December 2000, this retaining wall collapsed into the creek. This put two homes located near the stream and about 5 m (about 18 ft) above the creek bed in danger. The property next door also had a stone wall that was a concern. City leaders provided the money to rebuild the wall and protect the homes. About 107 m (350 ft) of stream bank was rebuilt with limestone boulders, plantings of native grasses and trees, and special soil reinforced with synthetic materials for strength. The completed project protected the three homes and made a beautiful, natural stream setting.

The collapse of the original stone wall made the stream bank vulnerable to erosion by water and gravity.

The completed project along Little Walnut Creek successfully reinforced the stream banks and prevented further erosion.

NOTES

Case Summary

3.5.1

Name: ____________________ Date: __________

Case name	*Case Study 4: Little Walnut Creek*
Case description (include problem)	*A retaining wall collapsed, one was threatened, and bank erosion occurred. Three homes were threatened by erosion.*
Case solution (describe and sketch)	*The solution was to rebuild the retaining walls with limestone boulders, rebuild the stream banks with reinforced soil, and plant native grasses and trees to hold the soil.* *Students' sketches will vary, but should clearly show the limestone boulders supporting the retaining wall, the reinforced soil supporting the stream bank, and vegetation.*
Reason(s) that solution was used	*The wall, reinforced banks, and vegetation prevented further erosion.*
Outcomes (expected, unexpected, good, problematic)	*The wall prevented soil from sliding downhill into the creek. The vegetation's roots helped hold the soil in place.*
Conclusion (What did we find out about erosion management?)	*Walls can help prevent erosion caused by water acting on soil banks. Vegetation can help reduce erosion by holding soil in place.*

NOTES

Case Summary 3.5.1

Name: ______________________ Date: ____________

Case name	*Case Study 5: Fort Branch Watershed*
Case description (include problem)	*A confined drainage channel with eroding banks.*
Case solution (describe and sketch)	*The solution was to rebuild the bank using logs made of shredded coconut husks, wrapped with rope webbing, and to control the slope with rocks.* *Students' sketches will vary, but should clearly show the logs and rocks supporting the banks.*
Reason(s) that solution was used	*The wall, reinforced banks, and the rocks helped reduce erosion by changing the slope.*
Outcomes (expected, unexpected, good, problematic)	*The walls and rocks prevented soil from sliding into the channel, and protected the banks from erosion.*
Conclusion (What did we find out about erosion management?)	*Walls can help prevent erosion caused by water acting on soil banks. The rocks are used to alter the slope.*

NOTES

Case Study 5: Fort Branch Watershed

On the Fort Branch Watershed at Woodmoor Drive, there is a confined drainage channel. The sides of the channel were cut down, and the banks were eroded. The town reconstructed the bank and created a winding, more natural-looking channel using logs made of compressed natural materials, such as shredded coconut husks, and wrapped with rope webbing. The slope of the channel was controlled using rocks.

You can see the damage from erosion at the Fort Branch Watershed before the channel was rebuilt.

Effective erosion-control methods created a natural-looking channel and prevented further erosion.

META NOTES

No conclusive evidence has proved that the steepness of a slope affects the rate of erosion, but it is implied in some of the case studies. As the slope steepness increases, erosion increases. You might give students an example of a loosely crumpled up piece of paper reaching the edge of your desk with nothing below, and the edge of your desk with a ramp (use a cardboard box or the like) attached. The paper reaching the cliff falls all the way down, the crumpled paper reaching the ramp falls down more slowly and may not make it all the way down the slope.

NOTES

Stop and Think

15 min.

Groups answer the Stop and Think *questions and present the case studies.*

Stop and Think

For each case study you read, answer the following questions:

- What erosion problem were they addressing?
- What erosion-control methods were used to address the problem?
- Why did they think the chosen method would be a good one to use?
- What happened?
- In what ways did the erosion control work as planned, and in what ways, if any, did it create new problems?

Discuss the answers in your group, and be prepared to share your answers with your classmates.

Reflect

Working with your group, identify ways of using some of the erosion-control methods you have just read about. Some of the *Reflect* questions will have you thinking about where you have seen these methods used. Others will help you think about what might be useful around the basketball court. Be prepared to discuss your answers with your class.

1. Which of the erosion-control methods you read about have you seen used in your neighborhood or community? How was each used? Why do you think each was used in that place? Draw a diagram to help you communicate how and why each was used.
2. For each of the erosion-control methods you have identified, which of them might be useful at the basketball court? Describe why and how it might be used. What do you think will result from using it? Think about good things and problems that might result. Draw a diagram to help you communicate.
3. Choose another erosion-control method you think might be useful around the basketball court. Describe how that erosion-control method is used in one of the case studies. Why do you think you could use this approach for the basketball court? How might you use it? What do you think will result from using it? Think about improvements and problems that might result. Draw a diagram to help you communicate how and why it might be used.

○ Get Going

Inform groups that the *Stop and Think* questions are the same as the questions on the *Case Summary* page, but worded differently. Let students know they will be presenting their responses to the class and ask them to prepare for their presentation. The presentations should be informal and brief without posters. Give students about five minutes to prepare.

△ Guide and Assess

Once groups have finished answering the questions, have them present their case studies using their answers to the *Stop and Think* questions or the *Case Summary* page.

As groups present, record important ideas on the board, and have students record the same information in their *Case Summary* pages.

At the end of all of the presentations, guide students to understanding how these case studies may help them with their challenge. Ask students what was similar about all the case studies. *(All involved banks eroded by water.)* Ask them what was similar and different about these studies and their challenge. *(The* Basketball-Court Challenge *involves a steep bank and the concern of water falling on top rather than a creek or stream.)*

⬡ Get Going

Have students discuss and answer the *Reflect* questions with their groups. Emphasize that they should be prepared to discuss their ideas with the class.

Tell the class they should describe two case studies for Question 3.

As groups discuss the *Reflect* questions, ask students how they can apply what they are learning to the *Basketball-Court Challenge.* When you update the *Project Board*, make sure that these ideas are discussed.

Hold a discussion eliciting students' responses to the *Reflect* questions. These questions transition discussion into updating the *Project Board.*

When groups have finished answering the *Reflect* questions, draw students' attention to the *Project Board.* Ask students what they think they know about erosion control. They can add this to Column 1. What do students think they need to learn more about? They can add this to Column 2.

TEACHER TALK

"Now that we've discussed these case studies, what do you think we know about erosion and erosion control? What were some of the examples of erosion control that you've seen first hand? What do we need to learn to address the *Basketball-Court Challenge?*"

Reflect

15 min.

Students answer the Reflect *questions.*

META NOTES

The *Reflect* questions should get students thinking about examples of erosion control they have seen first-hand and those they have read about. They should engage students in thinking about how to apply what they're learning to the ***Basketball-Court Challenge.***

Update the *Project Board*

5 min.

The class updates the Project Board.

META NOTES

From the case studies, students might say that controlling the slope is a method of erosion control. Since they were not actually provided with evidence of this, it should be listed as an item to investigate.

4. Are there any other erosion-control methods used in the case studies that might be useful for the basketball court? If so, answer the same questions about them.

Update the *Project Board*

Earlier, you began a *Project Board* centered on the idea of learning about what erosion is and how to manage it. Now you have read some case studies about how others have solved their erosion problems. You know more about the factors that cause erosion and different ways to stop it. You are now ready to fill in the *Project Board* more completely.

What is most important to add to the *Project Board* right now are your ideas about how you might control erosion in the *What do we think we know?* column. It is also important to add what you still need to find out about erosion control to address the challenge in the *What do we need to investigate?* column. Identify erosion-control methods you have read about. Then identify what else you need to know about each of those methods to design an erosion-control method for the basketball court.

The *Project Board* is a great place to start discussions. You may find that you disagree with other classmates about what you know about an erosion-control method. If so, put a question about it in the *What do we need to investigate?* column. Discussing disagreements is a part of what scientists do. Such discussions help scientists identify what they or others still do not understand well and what else they still need to investigate to understand more fully.

What's the Point?

As you read each case study, you found some similarities between these situations and the basketball-court situation. You were able to see how others solved erosion problems caused by water and gravity. Retaining walls, terracing, and drainage systems can be used to direct the flow of water and keep water flowing in places where it cannot cause damage. Planting native grasses and trees has also been used to control erosion. Plant roots anchor the soil and other materials, preventing them from being carried away. In the cases you read, erosion-control methods were combined to solve erosion problems.

Students may information to add in Columns 3 and 4 since they have learned about some erosion control methods from the case studies. They may want to describe retaining walls as a method of erosion control, or planting natural vegetation with roots to hold the soil in place.

Assessment Options

Targeted Concepts, Skills, and Nature of Science	How do I know if students got it?
Scientists often work together and then share their findings. Sharing findings makes new information available and helps scientists refine their ideas and build on others' ideas. When another person's or group's idea is used, credit needs to be given.	**ASK:** What did you learn from listening to other groups' presentations about their cases? How did this help you? **LISTEN:** Students should identify specific cases about what has worked and what has not worked to control erosion.
Identifying factors that could affect the results of an investigation is an important part of planning scientific research.	**ASK:** How were solutions to these erosion problems identified? **LISTEN:** Students should note one important step was to identify the factors that led to erosion and then to find a way to stabilize the movement of soil.
Erosion is the process of soil and other particles being displaced by water, waves, wind, and gravity.	**ASK:** What causes and effects of erosion did you see in these cases? **LISTEN:** Students should discuss the phenomena that they recorded in their *Case Summary* pages. For all the case studies, moving water caused the erosion problems.

Teacher Reflection Questions

- What troubles did students have analyzing the case studies? How did students relate the case studies to the *Basketball-Court Challenge?*
- What part of going through the first case study with the class was most beneficial and why?
- What difficulties arose in transitioning from the *Reflect* questions to the *Update the Project Board* segment? What ideas do you have for next time?

SECTION 3.6 INTRODUCTION

3.6 Plan

Model Erosion Control

1½ class periods ▶

A class period is considered to be one 40 to 50 minute class.

Overview

Each group picks, or is assigned, one of the erosion-control methods the class identified in *Section 3.5* to investigate. They design a model of the basketball-court site using stream table trays, soil, and other materials and plan a simulation of the erosion-control method. Then they share their designs and plans with the class, getting feedback from their classmates and identifying problems and areas for improvement in their models and simulations. Using the ideas they got from the presentations, they revise their models and plans, making sure that they are ready to run their simulations.

Targeted Concepts, Skills, and Nature of Science	Performance Expectations
Scientists often work together and then share their findings. Sharing findings makes new information available and helps scientists refine their ideas and build on others' ideas. When another person's or group's idea is used, credit needs to be given.	Students should use what they learned during presentations to revise and improve their models and simulation plans.
Criteria and constraints are important in design.	Students should identify the important features of the real world to replicate in their models.
Scientists must keep clear, accurate, and descriptive records of what they do so they can share their work with others and consider what they did, why they did it, and what they want to do next.	Students should record their model designs and simulation plans, along with the reasoning behind them in *Erosion-Control Model* pages and *Our Simulation* pages.
Identifying factors that could affect the results of an investigation is an important part of planning scientific research.	Students should identify features of the real world that could lead to variation in their simulations.

Targeted Concepts, Skills, and Nature of Science	Performance Expectations
Scientific investigations and measurements are considered reliable if the results are repeatable by other scientists using the same procedures.	Students should make simulation plans that are specific enough so the procedure is replicable and the results repeatable.
In a fair test only the manipulated (independent) variable, and the responding (dependent) variable change. All other variables are held constant.	Students should design simulations that are fair tests of the erosion-control methods they are investigating.
Erosion is the process of soil and other particles being displaced by water, waves, wind, and gravity.	Students should identify the effects of erosion that they want to control.
Scientists use models to simulate processes that happen too fast, too slow, on a scale that cannot be observed directly (either too small or too large), or that are too dangerous.	Students should design models and simulations that will allow them to find out how erosion will affect the basketball-court site.

Materials	
1 per group	Stream table tray Small plastic cup Spray bottle, filled with water Ruler Plastic drop cloth Drain bucket
1 bin per group	Potting soil Spanish moss
1 per student	Disposable gloves *Erosion Control Model* page *Our Simulation* page
1 per class	Paper towel roll

Activity Setup and Preparation

In this section, groups design an erosion-control model for one of four erosion control methods for the *Basketball-Court Challenge.* They will investigate their models in the next section. You may want to select erosion-control methods they listed on the *Project Board* or a method they bring up in the beginning of this section. Consider having them test out vegetation and retaining walls, or managing water flow and drains. There may be some overlap in methods tested.

Construct a model of the basketball court to show students and to test out the materials. We strongly suggest using native soil for the model. The Spanish moss can be used as vegetation. Tear it up so it can sit down on the soil. Students might use popsicle sticks to construct a wall. The drain hosing for the two groups testing drains should be fitted to the hole in the stream table and the hose should pour into a container. Straws could model piping to move the drain to other locations in the model. Models should represent the image shown in the student text and fit with the dimensions described.

Homework Options

Reflection

- **Science Process:** How confident are you of the prediction you made for your simulation? What else could happen besides what you predicted? *(This question is meant to engage students in thinking about what they expect to happen.)*
- **Science Process:** What do you think you will learn from running a simulation that you did not learn from the case studies? *(Students should recognize that the case studies told them what happened in a very specific environment and set of conditions. A simulation is required to see what might happen in a different environment.)*

Preparation for 3.7

- **Science Process:** What are some ways that the scale of your model will affect your simulation? How can you make sure the small scale does not affect your results? *(Groups will have to be careful that they scale down rain, winds, and any other atmospheric effects. Mist from the spray bottle, would be drizzie on a large scale.)*

SECTION 3.6 IMPLEMENTATION

◀ $1\frac{1}{2}$ class periods*

3.6 Plan

Model Erosion Control

You have just identified ways you think erosion can be controlled. It is now time to investigate how well each of these methods might work to control erosion around the basketball court. Each group will investigate a different erosion-control method by building a model of it and then running a simulation using your model. To do this, you will need to build a model of the basketball court site and the erosion-control method assigned to your group. Each group will receive a small container of soil to use in building their models. The soil in the container is similar to that surrounding the basketball court.

Architects usually build scaled-down models before the actual building begins. This way, they can get feedback on their designs and identify any potential problems.

Your first step will be to design your model so you can investigate the application of your erosion-control method to the basketball court site. Then you will plan your simulations. After discussing these with the class, you will build and test your erosion-control method.

Remember that your goal right now will be to learn how someone might use your assigned erosion-control method at the basketball court site. You will need to identify where to place it, how to place it, and what is likely to

3.6 Plan

Model Erosion Control

5 min.

Introduce the activity to the class.

○ Engage

Elicit from students their ideas about erosion control around the basketball court and tie it in to how they will be building a scaled-down model of the basketball site to test some erosion-control designs. Record students' ideas.

*A class period is considered to be one 40 to 50 minute class.

TEACHER TALK

"What are some erosion control methods you think would work for the basketball-court site? Why do you think this method would work?

You have read about some erosion-control methods, and we have listed some ideas you have for controlling erosion around the basketball site. What do you think we should do now?

In this section, you are going to build a scaled-down model of the basketball-court site to test out an erosion-control method. I will assign each group one method of erosion control to test. In the next section, you will get to test how well the erosion-control method works for your model."

NOTES

happen as a result of using it. You should aim to identify both its strengths and weaknesses when used around the basketball court. In a later section, you will have a chance to design what you think will be the best approach to managing erosion around the basketball court. That solution might combine several of the methods you are investigating in this section and the next.

Be a Scientist

Investigating with a Model

In the *Lava Flow Challenge*, you read about and discussed using models to help you investigate problems. Your class reviewed how a model makes it possible to test ideas and get scientific results when it is not very easy or even possible to study the real-life situation. For example, it is very difficult for scientists to complete investigations about galaxies millions of light-years away. They cannot travel there or manipulate objects in the galaxy. So scientists create models that are similar to the actual galaxy, and they investigate using those models.

Remember that models are representations of something in the real world. Simulations use models to imitate, or act out, real-life situations. You are going to investigate erosion-control methods by building a model of the basketball court and your erosion-control method. Then you will use your model to simulate the erosion-control method you are investigating. It would be too expensive to try out all the erosion-control methods at the real basketball court site. Your model and simulation will help you determine how effectively your erosion-control method will be at controlling erosion at the basketball court site and how to make it effectively do its job.

In the next few sections, each group will build a model that includes the basketball court, the hill above it, and one of the erosion-control methods. Using their model, each group will then simulate erosion control around the basketball court.

Remember that a simulation can only teach you something if it is run on a model that accurately represents the conditions of the real-life environment. Models can never match the real world exactly. For one thing, models are usually scaled down. "Scaled down" means smaller than the actual size. Models also do not have all the details of the real world. For example, the soil in your model will not include all of the kinds of particles that can be found in the soil on the real hill.

Be a Scientist: Investigating with a Model

5 min.

Students discuss how the model will be used to investigate erosion control.

△ Guide

Briefly remind students that scientists use models when they are studying something that happens too fast, too slow, on a scale that cannot be observed directly (either too small or too large), or that are too dangerous. Remind them of the lava flow model—that modeled something too dangerous and too large.

Elicit from the class their ideas of why it is best to run a simulation to test ideas using a model of the basketball court. Students should realize not only is the site too large to test directly, but the model will allow them to test out many ideas and how effective they are before actually disturbing the site. Emphasize that it is important to make significant features of the model as close as possible to the real phenomenon. With the class, review the image of the proposed court in the *Learning Set* Introduction in the student text, and the measurements. Remind students of the class list of criteria and constraints for the *Basketball-Court Challenge*.

NOTES

It is important, however, to design models so they match the real world in the most important ways. Any model you design should match the real-world factors you know are important in managing erosion. For example, you know that the type of material on Earth's surface affects erosion. You also know that the amount of moisture in the material is a factor in erosion. It will be important to decide what real-world conditions should be included in your model. Then design your model to match those real-world conditions as closely as possible.

Design Your Model

Each group will investigate a different erosion-control method by building a model of it on the basketball court site, and then running a simulation. With your group, make decisions together about how to design and build a model of the basketball court site and your erosion-control method. Each group will receive a small container of soil to build a model. The soil in the container is similar to that surrounding the basketball court.

Erosion Control Model 3.6.1

Name: ______ Date: ______

What we are modeling:

Our models' parts (include evidence and science knowledge)
Design decision | Reason

Similarities and Differences
How our model is like the real world:
How our model is different from the real world:

How we will build our model:
Description:

Sketches:

As you design and build your model, use these questions to guide your decision-making:

- What factors, other than soil type, affect erosion in the real world and should be included in the model?
- How will the scale of the model affect the simulation and your results?
- What constraints, identified earlier in the *Learning Set*, must be addressed in the model?

To model your group's erosion-control method, you will need to make many decisions:

- What materials you will use to model the erosion-control method?
- How will you construct any erosion-control devices you need to create, such as walls?
- Where on the basketball court site will you place those devices?

Design Your Model

15 min.

Groups design models to investigate erosion-control methods.

META NOTES

If any of the groups in your class have outstanding model designs or are having great difficulties designing a model, consider having groups give mini-presentations to show how they are designing their models.

△ Guide

Assign each group an erosion-control method and tell them they will need to design a model to investigate that method. They will build their models during the next section. The model will be of the basketball-court site using a bin with native soil, other materials such as Spanish moss to simulate vegetation on the hill, and should include their erosion-control method in their model.

Go over the questions and guidelines with the class. Have the materials and *Erosion-Control Model* pages available, and have groups plan their models. They should not build their models yet, but they should think about how they can use the moss, slate chips, gravel, and other materials to model the erosion-control methods they are investigating. Emphasize that students will need to think about how to model the basketball-court site and how to model the erosion-control method.

△ Guide and Assess

As groups are working on their models, check to see if the models are feasible, and make sure they record the decisions they make and the reasons in their *Erosion-Control Model* pages. Check to see that they have incorporated answers to the guiding questions in the student text. Use these questions to guide groups having difficulties starting.

If you see any features of a model that you think will not work for the simulation, note them and make sure that the group tests those features after they build their models. If any groups are struggling, get students from other groups to help them.

NOTES

- What changes, if any, do you need to make to the landscape around the basketball court to use your method?
- How will you attach the devices to the basketball court site?
- What things will you want to find out about your erosion-control method when you run your simulation?

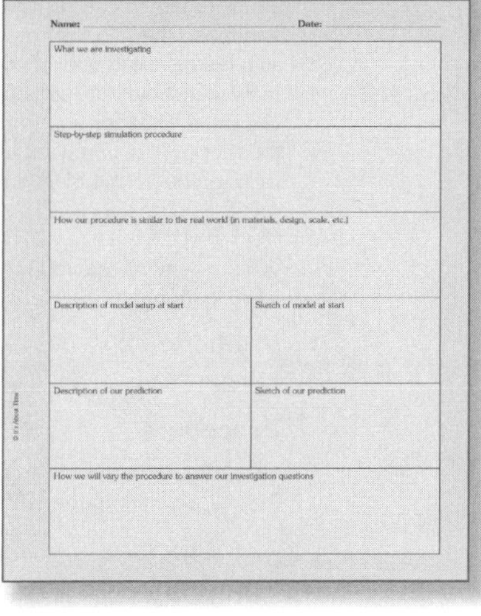
Our Simulation

Name: Date:

What we are investigating

Step-by-step simulation procedure

How our procedure is similar to the real world (in materials, design, scale, etc.)

Description of model setup at start | Sketch of model at start

Description of our prediction | Sketch of our prediction

How we will vary the procedure to answer our investigation questions

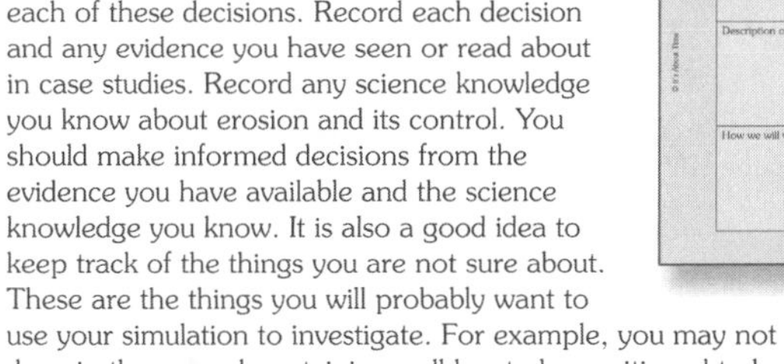

Record all of your design decisions, evidence for them, and answers to the questions above in the appropriate places on an *Erosion Control Model* page. You will need to carefully consider everything you have been learning as you make each of these decisions. Record each decision and any evidence you have seen or read about in case studies. Record any science knowledge you know about erosion and its control. You should make informed decisions from the evidence you have available and the science knowledge you know. It is also a good idea to keep track of the things you are not sure about. These are the things you will probably want to use your simulation to investigate. For example, you may not know how deep in the ground a retaining wall has to be positioned to be able to hold back the dirt behind it.

Plan Your Simulation

After designing your model, you will need to plan and design your simulation. You will be running your simulations to learn about how to make your erosion-control method work at the basketball court. You have already identified some questions you want to answer about your erosion-control method. You should design your simulations to help you answer those questions. Use *Our Simulation* pages to record your plan.

Remember that your simulations should re-create real-world conditions as closely as possible. Think about such things as how water should be applied to the model and how much water should be applied to the model. One of your decisions might be whether you will simulate a heavy rainfall or a light rain shower. If you want to find out how heavy a rain your simulation-

Plan Your Simulation

15 min.

Groups design procedures that will show how their erosion-control methods will be tested.

△ Guide

After groups have designed their models, go over the guidelines with students and have them plan their simulations. Emphasize that they will need to consider the different conditions that could exist at the basketball-court site, such as saturated or dry ground, and the different kinds of precipitation that might occur, such as heavy or light rainfall. Emphasize that they will need to write a detailed description of their procedures addressing all the bulleted points listed in the student text. Distribute to students *Our Simulation* pages and discuss how to use it. Let students know they will be presenting their results to the class.

control method can withstand, you might want to begin with a soft shower and make the rain fall harder and harder over time. You will run several simulations using your model. Think about how you want to start each one. How wet should the soil be, for example? How clean will the bottom of the box need to be for you to make good observations about erosion? Think, too, about how you want to start each simulation. Use the guidelines below to help you as you plan your simulations.

Questions

What questions are you investigating and trying to answer with this investigation?

Prediction

For each, what do you think the answer is, and why do you think that?

Procedure

Write detailed instructions for how to carry out your investigation. Include answers to the following questions:

- How will you set up the erosion-control model?
- How will you simulate erosion?
- How will you measure the model's performance and how will you record the data?
- How many trials will you run, and how will you set up your simulation again after each trial?
- What changes to your model will you have to make to answer your questions, and what different simulations will you need to do?
- What difficulties do you think you might run into when you build your model and run your simulations, and how will you address those?

You might need to set up and run your model several different ways to determine where to put your erosion-control devices and how to attach them to the ground. You will need to collect careful data to answer each of the questions you are investigating. There may be some difficulties collecting that data. It is always good to identify difficulties before you investigate so that you will know what to be especially careful about when you carry out your investigation.

☐ Assess

Check to see that groups have answered the bulleted items on page 91 and have reasonable procedures. If they do not, these are things to make sure are brought up during the communicate section if their fellow classmates do not bring it up during their presentations.

⬡ Get Going

Once groups have finished planning their simulations, have them present designs for their model and simulation using *Our Simulation* pages.

Your goal is to determine the effectiveness of your erosion-control method. You can use your results to identify the strengths and weaknesses of the erosion-control method you are investigating. Remember, that everybody will need your results to be able to address the challenge. When you present your model and plan for investigation to the class, make sure you can tell them why you think they can trust your results.

Use one *Our Simulation* page for each of the simulations you plan. Use the hints on the planning page as a guide. Be sure to write enough in each section so you will be able to present your simulation design to the class. The class will want to know that you have thought through all parts of your plan.

Communicate

To help you as you learn to design models and investigations, you will share your plan with the class. Others in the class have designed models and planned investigations to answer questions similar to the ones you are answering. You will probably see differences and similarities across these plans. In the class discussion, compare the plans. Notice similarities and differences. Identify the strengths of each plan. Think about what could be improved in each. Pay special attention to anything that might help improve your plans.

Revise Your Plan

With your group, revise your plans based on the discussion you just had in class. You might want to change the way you will build your model. You might want to be more specific about the way you run your simulations.

What's the Point?

You have just designed an investigation that will allow you to evaluate your erosion-control method for use at the basketball court site. In the past, you probably followed written steps to run an investigation. Here, you are designing an investigation yourself. Your big challenge is to discover how scientists work together to solve problems. One thing scientists do is collect data and use it as evidence. By designing your own investigation, you will experience how scientists do this.

Communicate

20 min.

Groups share and discuss their designs and simulation plans with the class.

Revise Your Plan

10 min.

Groups revise their plans based on what they learned from presentations.

△ Guide

As groups are presenting, encourage students to ask questions if they do not understand aspects of the model or the simulation plan, or if they think they see something that will not work. Model the kind of questions you expect students to ask by asking groups how their simulations approximate real conditions, how their models will show them if their erosion-control methods are likely to succeed or fail, and what they expect to happen when they run their simulation.

Emphasize that groups will have an opportunity to revise their plans, and that they should think about whether questions from the class reveal problems in their plans. They should also think about whether there are advantages to the way other groups have planned their models or simulations.

⬡ Get Going

Have groups revise their plans using what they learned from the presentations. This is a chance for them to change their model or their simulation, or to make their procedure more precise.

◇ Evaluate

Before the end of class, look at each group's plan and determine whether it needs more work. Simulation plans need to be specific enough so that they can be repeated. They need to specify how much of each material should be used, how much water should be used, how the materials should be landscaped, how the water should be used, and how the results should be evaluated.

Make sure that students are prepared to build their models and run their simulations immediately the next time class meets.

Teacher Reflection Questions

- What problems are you seeing in how students plan and conduct investigations? What steps will you have to take in the future to get them past these errors?
- How did students revise their models and/or simulation procedures? What evidence do you have that students see the value in sharing ideas and revising their own?
- Students need to present and discuss their investigation with their peers. How easily did you let students take control of the investigations? How much support did you need to provide them? What can be done to improve their peer-to-peer discourse?

NOTES

SECTION 3.7 INTRODUCTION

3.7 Investigate

2 class periods ▶

A class period is considered to be one 40 to 50 minute class.

Simulate Erosion Control

Overview

Students build models of the basketball-court site and test them, recording any changes they implement to replicate real-world conditions. Once their models accurately replicate important features of the basketball-court site, they run simulations to test their erosion-control methods. The erosion-control model will generally not work as expected the first time; students record what happened and revise the model to make it more effective, recording their changes and the reasons for their changes. Groups run the simulation again, iterating the process as many as three times. Then they share their results with the rest of the class, during an *Investigation Expo*. The class discusses what they have learned about erosion-control models and records this information on the *Project Board*.

Targeted Concepts, Skills, and Nature of Science	Performance Expectations
Scientists often work together and then share their findings. Sharing findings makes new information available and helps scientists refine their ideas and build on others' ideas. When another person's or group's idea is used, credit needs to be given.	Students should work with their groups to investigate the effectiveness of erosion-control methods and share their results with the class.
Scientists must keep clear, accurate, and descriptive records of what they do so they can share their work with others and consider what they did, why they did it, and what they want to do next.	Students should record test results for their models, results of the simulation, and revisions to their erosion-control models in *Running Our Erosion-Control Model* pages.
Scientific investigations and measurements are considered reliable if the results are repeatable by other scientists using the same procedures.	Students should use simulation procedures that are specific enough so that they and other students can replicate them.

Targeted Concepts, Skills, and Nature of Science	Performance Expectations
In a fair test only the manipulated (independent) variable, and the responding (dependent) variable change. All other variables are held constant.	Students should only modify the erosion-control method when they revise their models, holding the rest of the simulation constant.
Erosion is the process of soil and other particles being displaced by water, waves, wind, and gravity.	Students should observe how the soil and other materials in their models are displaced by water and wind.
Scientists use models to simulate processes that happen too fast, too slow, on a scale that cannot be observed directly (either too small or too large), or that are too dangerous.	Students should model the erosion that might take place at the site of the basketball court using small models that they can modify by hand in the classroom.
Scientists make claims (conclusions) based on evidence (trends in data) from reliable investigations.	Students should review the results of their investigations and make conclusions.

Materials	
1 per group	Stream table tray Small plastic cup Spray bottle, filled with water Ruler Plastic drop cloth Drain bucket Presentation materials
1 bin per group	Native soil Spanish moss Slate chips Landscape materials
1 per classroom	Paper towel roll
3 per student	*Running Our Erosion-Control Model* page
1 per student	Disposable gloves *Our Simulation* pages

Materials	
1 per class (optional)	Digital camera

Caution: Watch out for spills.

Activity Setup and Preparation

Consider bringing cameras to class to take photographs of the models at various stages during the simulations.

Before class begins, organize the materials and equipment to allow students to expeditiously run their investigations. Prepare containers of soil to distribute to groups. Students will need to drain water and rebuild the slope after each test. Have buckets ready for students to drain water and any saturated mixture.

Homework Options

Reflection

- **Science Process:** What changes did you make to your model as you repeated your simulation? Why did you make these changes? *(Students should have kept most features of their model the same from simulation to simulation. If they changed the model significantly after testing it without the erosion-control method in place, they may not know what caused the changes they observe.)*

Preparation for 3.8

- **Science Content:** Do you think the erosion-control method you investigated could be effective at the basketball-court site? Why? *(Students should support their answers with observations from their simulations.)*

NOTES

SECTION 3.7 IMPLEMENTATION

◀ *2 class periods**

3.7 Investigate

Simulate Erosion Control

Have your teacher check your plan before you conduct any investigation.

It is time to explore the effectiveness of your erosion-control method. You will do this by constructing your model and then simulating natural forces acting on the hill. As you simulate rain or wind acting on your hill you will be able to see how well your erosion-control method works. Remember, your goal is to investigate the effectiveness of your assigned erosion-control method.

Build Your Model

Your first step is to build your model. Use your revised design plan, and build your model the way you specified in your design.

Before running your investigation, you will need to make sure your model is built the way you intended.

- Does it look the way you want it to look? The size and placement of your hill and the basketball court should match the challenge. Your erosion-control device or method should look like those in the real world.
- Check that everything is in the places where you intend them to be.
- Check that everything is put together well. Anything that needs to be movable should be movable. Anything that needs to be held in place should stay in place well. Remember that moist soil holds things differently than dry soil. Make sure the conditions of the soil in your model are what you intended.

If everything is in place, then you are ready to see if your model will hold together in a simulation.

Test Your Model

Unless your simulation needs to begin with dry soil, you can test your model by gently simulating natural conditions. What happens to it when you apply a gentle rain? What happens with a gentle breeze? Remember that your model is a lot smaller than the real-world basketball court site. Take your

3.7 Investigate

Simulate Erosion Control

Build Your Model

15 min.

Groups build their models.

⬡ Get Going

Have groups begin building their models as soon as they come to class. Let them know that they will need to test their models and get your approval before they run their simulations.

☐ Assess

As groups are building their models, make sure they are using their designs. If they find that they have to modify their design, they should record the changes they made and the reasons for the changes.

*A class period is considered to be one 40 to 50 minute class.

Test Your Model

5 min.

Groups test their models.

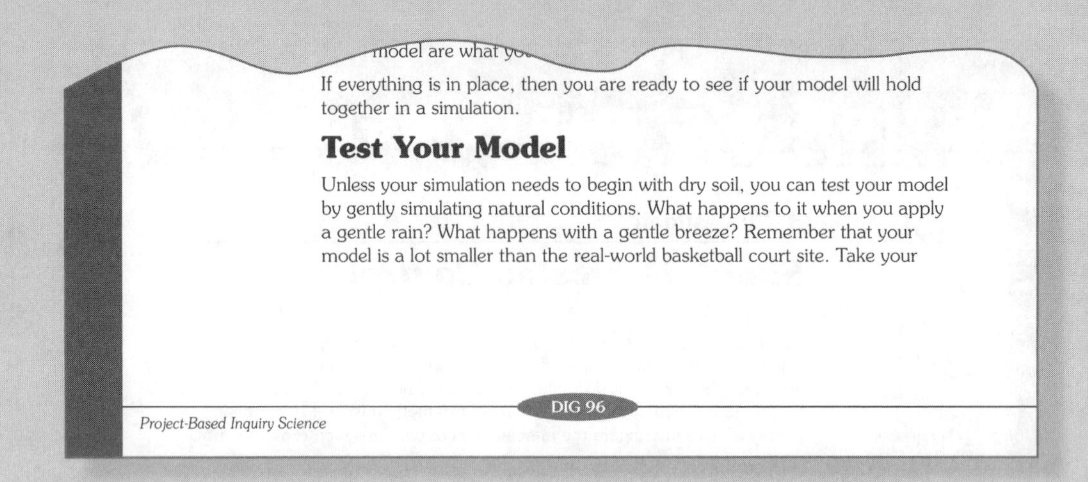

⬡ Get Going

Have students test the integrity of their models by saturating the soil with water and then simulating light and heavy rain. Their model should show a realistic level of erosion. If they have a landslide after a light misting, or if they have no erosion after dumping water on the model, then they will need to adjust it.

◇ Evaluate

Before groups run their simulations, visit each group and make sure their model meets all requirements. Test the integrity of the model by simulating soft rain (spraying from a spray bottle on the model).

NOTES

model's small size into account when you simulate rain. Think about the size of raindrops, direction of rain, and amount of rain as you decide how to apply a gentle rain to your model.

Try it out. Does everything hold together? If your model doesn't hold together in a gentle rain, it won't hold together in a harder rain. If it begins to fall apart, do what you need to do to make your structures stronger.

Run Your Simulation

Now that you have tested your model, you can use it to investigate the effectiveness of your erosion-control method. You have an investigation plan. It is time now to follow it. Record the results of each of your simulations on *Running our Erosion-Control Model* page. Make sure to record all the different kinds of data you specified in your investigation plan.

Your investigation plan may include changes you will make to the model and simulations using each variation. Make sure each *Running our Erosion-Control Model* page you use states which model you are using for your simulation.

You may also find that some of your simulations happen differently than you were expecting. Sometimes what you observe will give you ideas about other modeling and simulation you should do. For example, a retaining wall might fail in a way that gives you a new idea about where to place it on the hill.

In this kind of investigation, you can decide to change the model a little and run another simulation. Keep three important things in mind as you are doing this.

- Make sure every model and simulation is done for the purpose of answering one of your questions.
- Vary only one thing at a time when you change a model. That way, you can compare two instances of modeling to find out what caused different outcomes.
- Record every change you make to the model as you carry out your investigation. You should also record why you are making every change. To be able to draw conclusions, you will need to know what you did and why.

Run Your Simulation

20 min.

Groups run their simulations.

○ Get Going

When groups have finished building their models and you have checked them, distribute the *Running Our Erosion-Control Model* pages and have groups run their simulations. Emphasize they should follow the guidelines in the student text and record their results on the *Running Our Erosion-Control Model* pages.

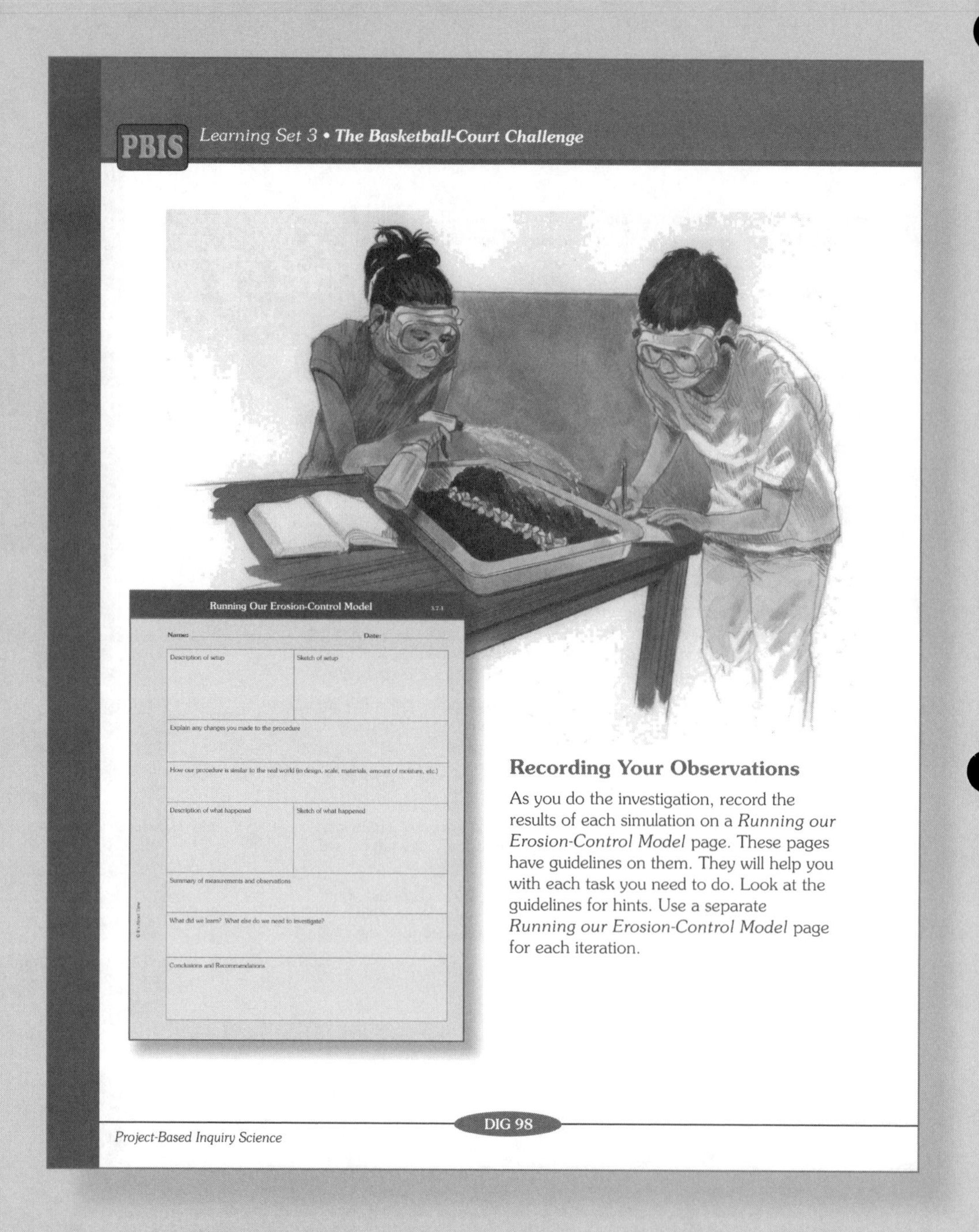

Running Our Erosion-Control Model 3.7.1

Name: Date:

Description of setup | Sketch of setup

Explain any changes you made to the procedure

How our procedure is similar to the real world (in design, scale, materials, amount of moisture, etc.)

Description of what happened | Sketch of what happened

Summary of measurements and observations

What did we learn? What else do we need to investigate?

Conclusions and Recommendations

Recording Your Observations

As you do the investigation, record the results of each simulation on a *Running our Erosion-Control Model* page. These pages have guidelines on them. They will help you with each task you need to do. Look at the guidelines for hints. Use a separate *Running our Erosion-Control Model* page for each iteration.

△ Guide and Assess

As groups run their investigations, they should carefully observe and record what happens in their *Running Our Erosion-Control Model* pages. With each erosion-control model, erosion will occur in unexpected ways (such as water running around the sides of a wall). Groups should fix the problems they discover with their erosion-control methods and record every change they make, with the reasons for making the changes. They should use a separate page for each revision, recording all of their observations and conclusions. Groups will need to do three iterations. Emphasize that they should use the same amount of water each time. They should keep all features of the model that they are not explicitly changing constant.

As groups run their simulations and revise their erosion control, monitor their progress and help them stay focused. If any groups are making multiple changes every time they revise their erosion-control model, encourage them to modify only one thing at a time. You may also see groups try to combine several different erosion-control methods. Groups should only combine methods if they have specific reasons for it. If groups experience landslides, they may be pouring too much water, or their model may be flawed.

Compare results from groups testing the same erosion-control method and note differences and similarities that you may want to bring out during the *Investigation Expo*. Note other things you think should be discussed during the *Investigation Expo*.

NOTES

Analyze Your Data

Review the data from your investigation. Look for any trends in your results. Use them as evidence to support your answers to the following questions:

1. How did you measure the effectiveness of your method? What happened when it was most effective? What happened when it was least effective?
2. Under what conditions was your erosion-control method effective at controlling erosion? How effective was it? Were there any new problems it caused when it effectively controlled erosion from the hill?
3. What evidence from your past investigations and readings did you use to design your model? Did you find that this evidence applied to this particular situation?
4. What challenge constraints affect the use of this erosion-control method?
5. What problems remain?
6. If you had the chance, what else would you investigate about this erosion-control method before deciding to use it at the basketball court site?

Communicate

Investigation Expo

When you finish your investigation, you will share your results with the class in an *Investigation Expo*. In this *Expo*, each group will take turns standing in front of the class and presenting their erosion-control models and investigation results. So others can learn from what you did and use your results when they design their best erosion-control method, you will make a poster that includes your design, the procedure you used to simulate erosion, and the meaning of your results.

In your presentation, you will explain to the class how your erosion-control model worked (or did not work). Include enough details in your presentation so your class understands how well your model worked to prevent erosion. Since each group tested a different erosion-control method, you need each other's information as you work toward the final model you will present to the school board. Answer the questions on the following page in your presentation.

Analyze Your Data

10 min.

Groups analyze their data using the questions in the student text.

META NOTES

A good place to end the first class period is at the conclusion of this segment.

Get Going

After groups have completed their investigations, have students use their *Running Our Erosion-Control Model* pages to answer the questions in *Analyze Your Data* segment. When they have answered them, they should briefly discuss their answers with their group and decide on their best answers. Let them know that they will need this information when preparing for their *Investigation Expo.*

☐ Assess

As groups discuss the *Analyze Your Data* questions, check to see if any of the questions are giving them trouble, consider bringing them up for discussion during the *Investigation Expo.* Look for the following in students' answers:

1. Students should have looked for displacements of material, changes in the shape of the slope, deposition at the base of the hill, and soil carried through the water into the runoff bucket.
2. Students should specify what kinds of precipitation their erosion-control method was able to withstand, as well as any other relevant conditions.
3. Students should have used the evidence from case studies involving the erosion-control method. They may use evidence about erosion from the earlier case studies. They may include how the Dust Bowl case study provided evidence that vegetation can help to reduce erosion.
4. Some of the methods students investigated may have been affected by the need to not interfere with surrounding properties.
5. Student should honestly evaluate their solutions and look for areas where they can improve them.
6. Answers to this question might address areas where students would like to improve their solutions, or they might point to ways to increase students' confidence in their results.

Communicate: *Investigation Expo*

30 min.

Groups present the results of their investigations during an Investigation Expo.

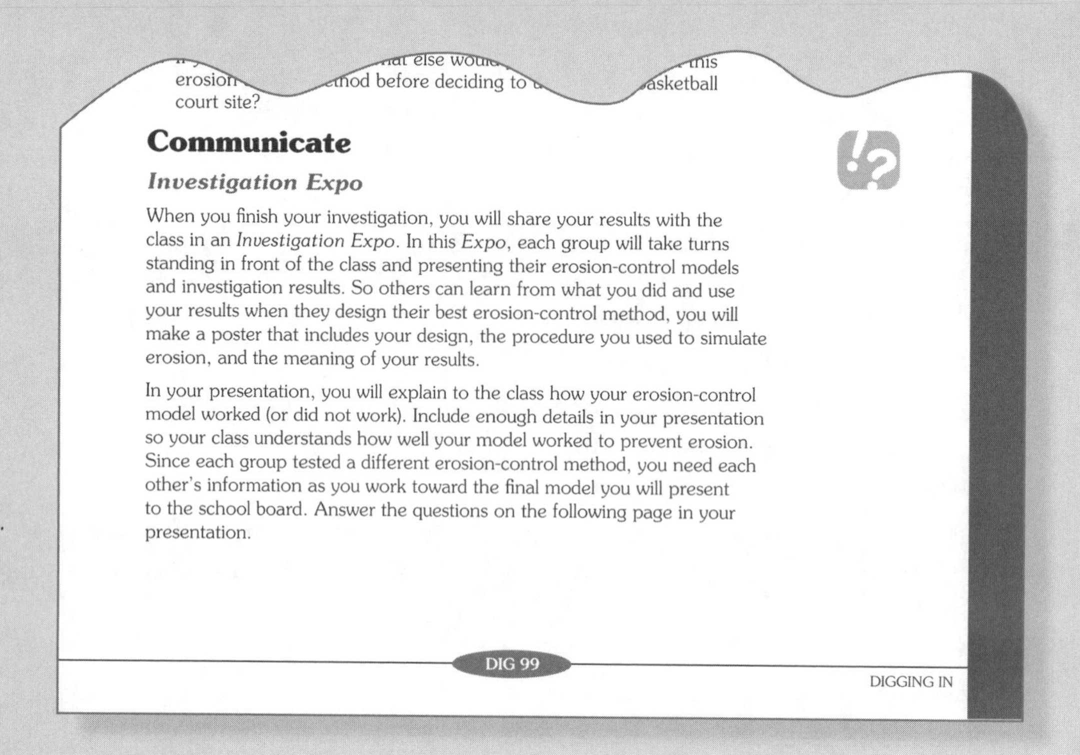

erosion ... method before deciding to ... basketball court site?

Communicate

Investigation Expo

When you finish your investigation, you will share your results with the class in an *Investigation Expo*. In this *Expo*, each group will take turns standing in front of the class and presenting their erosion-control models and investigation results. So others can learn from what you did and use your results when they design their best erosion-control method, you will make a poster that includes your design, the procedure you used to simulate erosion, and the meaning of your results.

In your presentation, you will explain to the class how your erosion-control model worked (or did not work). Include enough details in your presentation so your class understands how well your model worked to prevent erosion. Since each group tested a different erosion-control method, you need each other's information as you work toward the final model you will present to the school board. Answer the questions on the following page in your presentation.

DIG 99

DIGGING IN

△ Guide

Once groups have finished analyzing their data, discuss what is required for posters and presentations. Groups' presentations and posters should show what method(s) they tested, how they simulated erosion, how the method(s) worked or failed, what changes they made, and what they learned. Emphasize that students should address all the bulleted points in the student text as guidelines.

Emphasize that it is important for the class to understand how well the erosion-control method worked. If a group found that their erosion-control method was not effective, they should share this with the class. If they cannot honestly say that their investigation demonstrated whether the erosion-control method was effective or not, they should tell the class that the results were inconclusive and why.

Distribute poster materials and have groups create posters and prepare presentations. As groups are working on their posters and preparing for their presentations assist them as needed. Some issues that may arise are listed below.

- Students might have trouble deciding how to determine the confidence in their results. They can base this on how carefully they simulated real conditions and whether they controlled variables. Ask if all features of their model besides the erosion-control method the same when they ran the simulation as when they tested the model?

PBIS Learning Set 3 • The Basketball-Court Challenge

- What were you trying to find out from your investigation? What was your plan for answering those questions?
- What did you do to simulate erosion? Where did you pour the water onto your model? How much water did you pour? How did you make sure the way you poured the water on your model was similar to real rain? What changes, if any, did you have to make to your earlier procedure to answer your questions?
- How well did your erosion-control method work to manage the erosion? What was the path of the running water? How much soil was carried downhill by the water?
- How, exactly, did your erosion-control method work to manage the erosion? Describe exactly what happened in each trial.
- What changes did you make to your design to improve its performance? What evidence did you use to support those changes?
- How good do you think your model was at modeling your erosion-control method? How much do you trust your results?
- What did you learn about making this erosion-control method effective?

You might find out after running your investigation that your method did not work well. If so, you need to make sure your class knows why this is not a good method to use at the basketball court site.

You might have found that you did not construct your model well enough to learn about your erosion-control method. If that happened, be honest about it. Then explain how you might change the design to make it work better the next time.

As you listen to the investigation presentations of other groups, observe how their different erosion-control methods worked. In what ways is each one good at managing erosion at the basketball court site? What new problems does each one cause?

DIG 100

Project-Based Inquiry Science

- Students might include opinions in their interpretations or claims. Ask them if their conclusion is directly supported by their observations.
- Students might be using science knowledge incorrectly or incorrect information that they think is accepted science knowledge. Ask them what evidence from the case studies they can use.

Remind groups that everything they show on their poster or say in their presentations will have to be clear so that others can understand and follow their thinking.

⬡ Get Going

When students have finished their posters, have each group display their poster around the classroom. Have each group visit each poster for a minute or two. After all the posters have been viewed, hold the class presentations.

△ Guide

Have groups who investigated the same erosion-control method present their results in sequence. Have groups discuss similarities and differences in their results and what seems to be reliable for a given erosion-control method.

Emphasize to students in the audience they should think about which of the erosion-control methods would be useful at the basketball-court site and what problems they might cause.

Encourage students to ask presenting groups questions about how confident they are of their results and how the erosion-control methods they investigated might work at the actual basketball-court site.

Follow the presentations with a brief discussion. Are any of the results surprising? Which of the erosion-control methods were most effective?

NOTES

Reflect

Think about what you learned in the *Investigation Expo* and how well your model worked or didn't work. Answer the following questions and be prepared to discuss the answers with your class:

1. How was your erosion-control model supposed to work?
2. Did it work in the way you thought it would?
3. In what ways were you satisfied with the results?
4. In what ways were you dissatisfied with the results?
5. Given what you read in the case studies and what you saw in your investigations, what new things have you learned? What is your evidence?
6. How are the results of your investigation and the results from other groups going to influence your solution to the challenge? Why?

Update the *Project Board*

You have learned a lot now about how different erosion-control methods work or do not work, and why. It is a good time to go back to the *Project Board* and update it. You can add information in the *What are we learning?* column. Be sure to add the evidence that supports what you have learned in the *What is our evidence?* column.

What's the Point?

You have just run a set of simulations to test the effectiveness of a set of erosion-control methods. You were able to find out many things about how to manage erosion at the basketball court site. That is the point of modeling and simulating real-life situations. It would be far too costly to try out all the different erosion-control methods at the actual site of the proposed basketball court. By using models and running simulations, you can find out what works and what does not work before spending time and money on building an erosion-control system that may not solve your problem.

Reflect

15 min.

Student answer and discuss the Reflect *questions.*

Update the *Project Board*

10 min.

Update the Project Board *with the class.*

○ Get Going

Give students a few minutes to answer the *Reflect* questions with their groups.

□ Discuss and Assess

When groups have finished answering the *Reflect* questions, lead a class discussion of groups' answers. Listen for things that students learned, students disagree upon, and things that surprised students during the discussion. These might point to information that should be included on the *Project Board*.

△ Guide

Draw students' attention to the *Project Board*. Ask if any of the things they just discussed should go in the third column of the *Project Board (What are we learning?)*. Ask what conclusions from their investigations they can put in the third column. Ask for evidence to put in the fourth column. They should draw their evidence from their investigations and their classmates' investigations. Update the *Project Board* as students suggest things.

Update the second column with investigative questions for claims students disagree upon.

◇ Evaluate

All conclusions for the investigations should be placed in the third column with the supporting evidence in the fourth column.

Assessment Options

Targeted Concepts, Skills, and Nature of Science	How do I know if students got it?
Scientists often work together and then share their findings. Sharing findings makes new information available and helps scientists refine their ideas and build on others' ideas. When another person's or group's idea is used, credit needs to be given.	**ASK:** What did you learn about erosion-control methods from other groups' presentations? **LISTEN:** Students' answers should include the results other groups obtained through their investigations.
Scientists must keep clear, accurate, and descriptive records of what they do so they can share their work with others and consider what they did, why they did it, and what they want to do next.	**ASK:** How did you use the records you kept of your simulations and the results? **LISTEN:** Students should have used these records to prepare their presentations.
Scientific investigations and measurements are considered reliable if the results are repeatable by other scientists using the same procedures.	**ASK:** How confident are you that your results would be repeatable by other researchers? Why? **LISTEN:** Students should base their confidence on how well they controlled variables and how specific their procedures were.

Targeted Concepts, Skills, and Nature of Science	How do I know if students got it?
In a fair test only the manipulated (independent) variable, and the responding (dependent) variable change. All other variables are held constant.	**ASK:** What variables did you hold constant? **LISTEN:** Groups should have used the same amount of water and shape of the basketball-court site the same for each test.
Erosion is the process of soil and other particles being displaced by water, waves, wind, and gravity.	**ASK:** What kinds of erosion did you see in your simulation? What caused the erosion? **LISTEN:** Students should apply what they learned about the types of erosion from the case studies in *Section 3.5* to their simulations. Students should have identified running water or wind as a cause of the erosion.
Scientists use models to simulate processes that happen too fast, too slow, on a scale that cannot be observed directly (either too small or too large), or that are too dangerous.	**ASK:** Why was it useful to run a simulation to investigate erosion-control methods? **LISTEN:** Students should recognize here would be no way to test the erosion-control methods in full scale without risking serious damage to the basketball-court site and spending a lot of money.
Scientists make claims (conclusions) based on evidence (trends in data) from reliable investigations.	**ASK:** What conclusions were made from these investigations? What was the evidence? **LISTEN:** Students should have made conclusions about the effectiveness about the erosion-control methods they studied. Their evidence should be drawn from their observations.

Teacher Reflection Questions

- In what ways did you see students using what they learned from the case studies in their simulations? How can you help students connect the information they learned throughout this *Learning Set?*
- What difficulties did students have improving their erosion-control models? What can you do to assist students with these difficulties?
- How were you able to keep students focused on revising their models and obtaining the best results possible? What might you do differently next time?

NOTES

SECTION 3.8 INTRODUCTION

3.8 Recommend

Which Erosion-Control Methods Might Be Appropriate for the Basketball-Court Challenge?

◀ ***1 class period***

A class period is considered to be one 40 to 50 minute class.

Overview

Students learn how to construct a recommendation and develop their first recommendation for how to control erosion at the basketball-court site based on their investigations and analysis of case studies. They support their recommendations with evidence and science knowledge in the same way they constructed explanations. They share their recommendations with the class, and record the class's recommendations on the class *Project Board*.

Targeted Concepts, Skills, and Nature of Science	Performance Expectations
Scientists often work together and then share their findings. Sharing findings makes new information available and helps scientists refine their ideas and build on others' ideas. When another person's or group's idea is used, credit needs to be given.	Students should work with their groups to create recommendations and share their recommendations with the class.
Scientists must keep clear, accurate, and descriptive records of what they do so they can share their work with others and consider what they did, why they did it, and what they want to do next.	Students should use the results of their investigations recorded on the *Project Board* to support their recommendations, and they should record their recommendations on the *Project Board*.
Erosion is the process of soil and other particles being displaced by water, waves, wind, and gravity.	Students should use what they know about erosion to support their recommendations.
Explanations are claims supported by evidence, accepted ideas, and facts.	Students should support their recommendations with evidence, accepted ideas, and facts using *Create Your Explanation* pages.

Materials	
1 per student	*Create Your Explanation* page
1 per class	Class *Project Board*

Homework Options

Reflection

- **Science Content:** Write a report with the class's recommendations for how to control erosion at the basketball-court site to the school board. Make your report as persuasive as you can. *(Students should use evidence and science knowledge to support their recommendations.)*
- **Science Content:** Construct a recommendation using the evidence and science knowledge below.

  Evidence: Whenever we plant grass seeds in topsoil without covering it with dried out grass, the topsoil and seeds get washed away in the first rain.

  Science Knowledge: Bare topsoil erodes easily. Topsoil covered with vegetation does not erode as easily.

Preparation for 3.9

- **Science Process:** What do you think you need to do before you can give the school board a useful answer? *(Students have disjointed recommendations about the use of different erosion-control methods at the site, but they have not determined how these could work together. Students need to design a complete solution to the* Basketball-Court Challenge.*)*

NOTES

SECTION 3.8 IMPLEMENTATION

◀ *1 class period*

3.8 Recommend

Which Erosion-Control Methods Might be Appropriate for the Basketball-Court Challenge?

You now have some ideas about how well your erosion-control method works and how to make it work effectively. Perhaps you found that it is not possible to make it work effectively. Whatever you found in your investigation, you should now be ready to make a recommendation to your class about the use of your erosion-control method at the basketball court site.

A recommendation is a special kind of claim where you make a statement about what someone should do. The best recommendations also have evidence and science knowledge associated with them. To make sure your recommendations are good ones, and so you can convince your classmates, you will first state your recommendation as a claim. You will identify the evidence and science knowledge that supports it. You will then create an explanation that supports your recommendation.

Create Your Recommendation

With your group, write a recommendation about the use of your erosion-control method at the basketball court. You might want to write your recommendation in a form that suggests what a person can expect in a situation or what they might do. The *Making Recommendations* box has several examples showing how this can be done well.

Be a Scientist

Making Recommendations

A recommendation is a kind of claim that suggests what to do when certain kinds of situations occur. It can have this form:

When some situations occur, ***do*** or ***try*** or ***expect*** something.

For example, if you want to make a recommendation for crossing the street you might say the following:

Project-Based Inquiry Science DIG 102

3.8 Recommend

Which Erosion-Control Methods Might Be Appropriate for the Basketball-Court Challenge?

5 min.

Introduce what students will be doing.

○ Engage

Briefly elicit students' ideas about what they think would be a good erosion-control method for the basketball court and why.

TEACHER TALK

"You've tested an erosion-control method and shared your results. If you had to pick one method to recommend for the basketball-court site, what would you pick and why? Remember that your reasoning will have to convince the school board."

Create Your Recommendation

10 min.

Students learn what a recommendation is and then create recommendations for the basketball-court site.

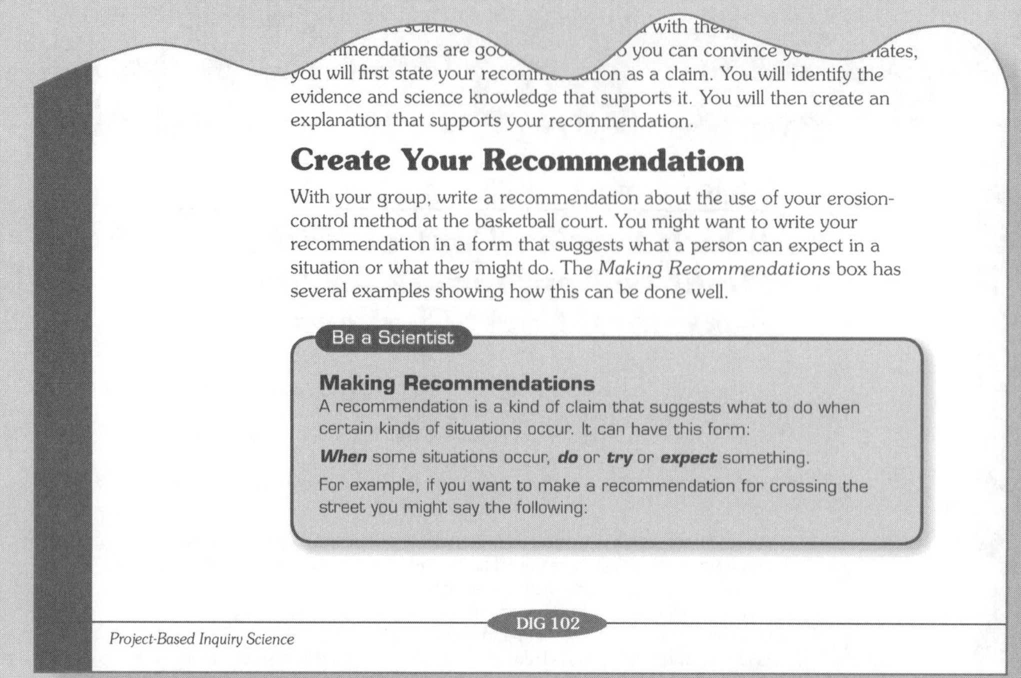

...mendations are goo... ...o you can convince yo... ...ates, you will first state your recomm...ation as a claim. You will identify the evidence and science knowledge that supports it. You will then create an explanation that supports your recommendation.

Create Your Recommendation

With your group, write a recommendation about the use of your erosion-control method at the basketball court. You might want to write your recommendation in a form that suggests what a person can expect in a situation or what they might do. The *Making Recommendations* box has several examples showing how this can be done well.

Be a Scientist

Making Recommendations

A recommendation is a kind of claim that suggests what to do when certain kinds of situations occur. It can have this form:

When some situations occur, ***do*** or ***try*** or ***expect*** something.

For example, if you want to make a recommendation for crossing the street you might say the following:

Project-Based Inquiry Science

DIG 102

△ Guide

Let students know each group will create recommendations either for or against the erosion-control method they investigated and present it to the class. Describe for students what a recommendation is.

TEACHER TALK

“A recommendations is a kind of claim that takes the form: *When some situation occurs, then do or try or expect something*. Sometimes recommendations can be in the form: *If something, then something.*”

Discuss the examples provided in the student text and emphasize that the recommendation has to be supported by evidence and science knowledge.

⬡ Get Going

Have students collaborate with their groups to come with their best group recommendation. They should start by writing a claim they agree on based on their investigation. Remind students that they may recommend for or against the erosion-control method they investigated.

When you have the right of way, ***expect*** that some cars will not be able to stop in time.

When you have the right of way, **look** both ways to make sure the traffic has stopped.

Recommendations might also begin with "if." For example,

If you have the right of way, and the traffic has stopped, ***then*** you can cross the street.

Support Your Recommendation

Remember that your recommendation should be written so it will help someone else. The school board members will be especially interested in knowing how you came up with your recommendation. Someone reading your recommendation should be able to apply what you have learned about erosion management and your erosion-control method.

The most trustworthy recommendations (ones people will follow) are those supported by evidence and science knowledge. Use the hints on a *Create Your Explanation* page to make your first attempt at supporting your recommendation. Your evidence will come from the experiments you ran earlier and the modeling and simulation you just finished doing. You can also use science knowledge from the cases you read.

Support Your Recommendation

10 min.

Students support their recommendations with evidence and science knowledge, just as they did with explanations.

△ Guide

Let students know that the most reliable recommendations are supported by evidence and science knowledge. Review how to create an explanation, and discuss how supporting a recommendation is similar to creating an explanation. When you create a recommendation, you connect a claim to evidence and science knowledge using logical statements, just as you do with explanations.

Communicate: Share Your Recommendation

10 min.

Update the *Project Board*

10 min.

The class records their recommendations and connects them to evidence and science knowledge.

After you have identified the evidence and science knowledge that support your recommendation, write an explanation statement to go with your recommendation that would help someone know why they should trust your recommendation.

Communicate

Share Your Recommendation

When you are finished, you will share your recommendation and explanation with the class. Make sure you trust each recommendation your class members present. If you don't understand the explanation that goes with a recommendation, ask questions. You will want to use your classmates' recommendations, and you can do that only if you believe they are trustworthy. So don't be shy about asking questions, do be respectful.

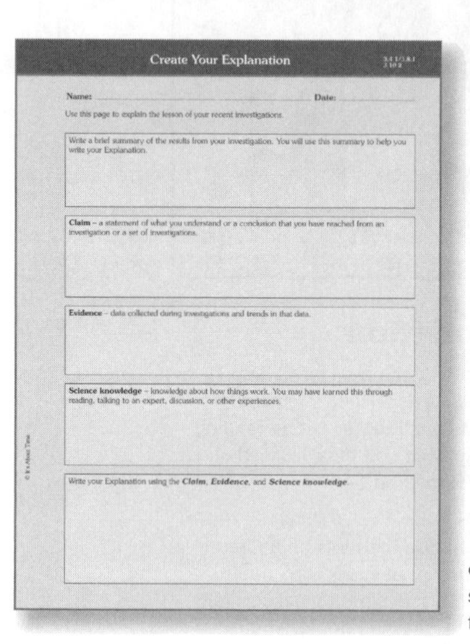

Create Your Explanation

Name: Date:

Use this page to explain the lesson of your recent investigations.

Write a brief summary of the results from your investigation. You will use this summary to help you write your Explanation.

Claim – a statement of what you understand or a conclusion that you have reached from an investigation or a set of investigations.

Evidence – data collected during investigations and trends in that data.

Science knowledge – knowledge about how things work. You may have learned this through reading, talking to an expert, discussion, or other experiences.

Write your Explanation using the Claim, Evidence, and Science knowledge.

Update the *Project Board*

Your *Big Challenge* is to design an erosion-control method to be used around a proposed basketball court site. The last column on the *Project Board*, *What does it mean for the challenge or question?*, is the place to record recommendations about how to address the challenge. You can now add in this column the recommendations each group wrote and presented for their erosion-control method. In addition, draw lines to the evidence and science knowledge (in the third and fourth columns) that support each of the recommendations. That way, you will be able to keep track of the reasons for each recommendation. When you write up your solution to the challenge, you will want to identify for the school board the evidence you collected that supports your solution.

⬡ Get Going

Distribute the *Create Your Explanation* pages and have groups use them to support their recommendations with evidence from their investigations and science knowledge from their previous studies. Remind students the evidence they need is in the fourth column of the *Project Board* and comes from the investigations they did in *Section 3.3* studying erosion and their investigation studying erosion control in *Section 3.7*.

What's the Point?

Science is about understanding the world around you. Scientists learn about the world by doing investigations. They make claims and provide explanations based on evidence and the science they already know. Then they communicate their results to other scientists. Some claims are in the form of recommendations. A recommendation suggests ways of accomplishing goals. The most trustworthy recommendations are supported by evidence and science knowledge.

Throughout this school year, you will investigate a variety of phenomena. You will apply what you learn to answering *Big Questions* and achieving *Big Challenges*. Many times during each Unit, you will be asked to explain your understandings and knowledge. You will write and share your understanding in explanations. Then you will make recommendations from the explanations you created. To support your recommendations, you will use the same evidence and science knowledge that helps you make claims. As you move through each Unit, you will have many opportunities to edit and improve your explanations and recommendations.

Working as a team, four students have done investigations, made claims, and formulated possible solutions. Now they are collaborating on plans to test their solutions for the Basketball-Court Challenge.

⬡ Get Going

Have each group present their recommendation, including the evidence and science knowledge they used to support it. Encourage students in the audience to comment and ask questions of presenting groups. Tell students they should ask for clarification as needed.

After all the presentations are completed, display each recommendation in a place where the class may edit it before it is placed on the *Project Board*. Group similar recommendations together, so that they may be combined, before writing on the *Project Board*.

△ Guide

Help the class make recommendations based on specific ideas by asking students to support their recommendations with evidence from the fourth column and science knowledge from the third column. Write the recommendations the class makes in the fifth column and draw lines to the evidence and science knowledge in the third and fourth column that support them. Let students know that they will use all of the recommendations to design a solution to the *Basketball-Court Challenge.*

After the class *Project Board* has been updated, remind students to update their *Project Board* pages.

Assessment Options

Targeted Concepts, Skills, and Nature of Science	How do I know if students got it?
Scientists often work together and then share their findings. Sharing findings makes new information available and helps scientists refine their ideas and build on others' ideas. When another person's or group's idea is used, credit needs to be given.	**ASK:** How was it useful to have groups create recommendations separately before discussing them as a class? **LISTEN:** Students should see that by creating explanations separately, groups were able to develop ideas based on their own investigations and the discussion that followed was more thorough than it might have been otherwise.
Scientists must keep clear, accurate, and descriptive records of what they do so they can share their work with others and consider what they did, why they did it, and what they want to do next.	**ASK:** How did you use your records to create your recommendations? **LISTEN:** Students should have drawn evidence from the records of their investigation and from the *Project Board*, and they should have used science knowledge from the *Project Board*.
Erosion is the process of soil and other particles being displaced by water, waves, wind, and gravity.	**ASK:** What were the causes of erosion that we investigated? **LISTEN:** Students should recall how the class investigated how water, gravity, and possibly wind caused erosion.

Targeted Concepts, Skills, and Nature of Science	How do I know if students got it?
Explanations are claims supported by evidence, accepted ideas and facts.	**ASK:** How did you support your recommendations? **LISTEN:** Students should have used evidence from their investigations and science knowledge to support their recommendations.

Teacher Reflection Questions

- What difficulties do you think students might have designing solutions based on their recommendations? What ideas do you have to help them with those difficulties?
- When engaging students, it is important to make connections with their experiences. What experiences might you use to make connections with? What experiences can you make connections to from the erosion walk? What types of erosion issues are occurring in the school district's community?
- Students need to present and discuss their recommendations with their peers. How easily did you let students take control of the presentations and discussion of recommendations? How much support did you need to provide them? What can be done to improve their peer-to-peer discourse?

NOTES

SECTION 3.9 INTRODUCTION

$1\frac{1}{2}$ ***class periods*** ▶

A class period is considered to be one 40 to 50 minute class.

3.9 Plan

Plan Your Basketball-Court Solution

Overview

Students plan a design to solve the *Basketball-Court Challenge*, using the class's recommendations to guide their decisions. As they plan, they sketch their designs and record the reasons for the decisions they make. They share their designs with the class and use feedback from the class to revise their plans. In the next section, students will build and test their designs.

Targeted Concepts, Skills, and Nature of Science	Performance Expectations
Scientists often work together and then share their findings. Sharing findings makes new information available and helps scientists refine their ideas and build on others' ideas. When another person's or group's idea is used, credit needs to be given.	Students should work with their groups to design solutions to the *Basketball-Court Challenge* and share their designs with the class.
Scientists must keep clear, accurate, and descriptive records of what they do so they can share their work with others and consider what they did, why they did it, and what they want to do next.	Students should sketch their designs and record design decisions and the reasons for the decisions.
Erosion is the process of soil and other particles being displaced by water, waves, wind, and gravity.	Students should consider the causes of erosion and the possible effects on the basketball-court site as they design their solutions to control erosion.

Materials	
1 per student	*Our Design Plan* page

Homework Options

Reflection

- **Science Process:** Did any of the questions or suggestions from the class surprise you? Summarize what you learned from them. *(Students' answers should draw on feedback from their* Plan Briefings.*)*
- **Science Process:** Do you think the design process would be more or less effective if everyone in the class was in one group? Why? *(Students' should recognize that getting feedback from other groups was an important part of the design process.)*

Preparation for 3.10

- **Science Content and Process:** What lessons from previous investigations do you think you can apply to testing your solution? Why? *(Students' answers should draw on experiences of running fair trials, recording results, and evaluating data, as well as using their science knowledge and evidence when they design how they will test their solution.)*

NOTES

$1\frac{1}{2}$ class periods* ▶

SECTION 3.9 IMPLEMENTATION

3.9 Plan

Plan Your Basketball-Court Solution

5 min.

Students begin to prepare their recommendations for the school board.

3.9 Plan

Plan Your Basketball-Court Solution

You are about to begin designing and constructing your solution to the *Basketball-Court Challenge*. Your solution can combine as many or as few erosion-control methods as your group feels is necessary. You have learned a lot about how each erosion-control method might work in your situation. You know you have to figure out how to control the water running downhill and the amount of soil and other materials it might carry onto the basketball court.

Using all the knowledge and evidence you gathered throughout this activity, you and your group will create a plan, or blueprint, for your solution. You will present it to your class in a *Plan Briefing*. After that, you will get a chance to build and test a model of your solution and revise your plan.

Design Your Solution

Using everything you have learned, work with your group to design an erosion-control solution for the basketball court. Remember that there are two houses to the sides of the basketball court. The school board says that the two houses and their lots must not be harmed by water or eroding soil.

Some of the erosion-control approaches you have investigated are good at directing water. Others are good at keeping soil in place. Some are not good at all. Your solution should take into account what you have learned about erosion and different erosion-control methods. Remember that the school board is more likely to approve a plan supported by evidence and science knowledge than one that is simply a good idea. As you design your solution, make sure you take into account everything you have been learning.

Using an *Our Design Plan* page, record your design. You will find space for diagrams of your design. You will also find a chart with space for you to list your design decisions and the reasons for each. It has three columns: one for design decisions, one for evidence that led you to make that decision, and one for science knowledge that supports it. You won't have evidence and science for every decision you make, but you need to have one or the other for each.

○ Engage

Let students know that they will be planning their design solutions for the *Basketball-Court Challenge* and they will be able to use multiple erosion-control methods in their design.

TEACHER TALK

"You have written many recommendations on the class *Project Board* that are supported by data and science knowledge you have learned. Do you think your recommendations are ready to be given to the school board? What more do you need to do? What do you think a professional contractor might do before sending off a recommendation for designing the basketball-court site?"

Let students know that they will be presenting their plans to the class and will have an opportunity to revise their plans before they actually begin building their design.

△ Guide

Let students know that they will be using everything they have learned. They should use their recommendations from the class to guide their design decisions. They should use the class *Project Board* and the class's criteria and constraints when planning their design. Consider reviewing the criteria of the *Basketball-Court Challenge* with the class before the class begins planning their designs.

Design Your Solution

20 min.

Groups design solutions to the Basketball-Court Challenge.

TEACHER TALK

"You want to plan a design that will meet all the criteria and constraints for the *Basketball-Court Challenge* using all you know about erosion and erosion control before you make your final recommendation to the school board. Like we have done previously, you will be sharing your plan with the class and have the opportunity to revise it after hearing everyone's plans. This way you can use others' advice and build on other's ideas."

⬡ Get Going

Distribute the *Our Design Plan* pages and emphasize that students should record all of their design decisions and the reasons for them in the chart. Then have groups get started.

△ Guide and Assess

As groups are working on their designs, ask students what ideas they have discussed. Ask how they are using the class's recommendations to guide their decisions. Check to see that the science knowledge and evidence support their idea. Ask for clarification if anything is not clear. Check if students are including opinions and guide them to understand that opinions are not part of evidence or science knowledge.

Communicate Your Plan: Plan Briefing

40 min.

Introduce Plan Briefings *and lead the class in preparing, giving, and discussing their design plans.*

Most of your design decisions will be based on recommendations made by different groups. Each recommendation has evidence and science knowledge associated with it. If you use the *Project Board* as a resource as you design your solution, it should be easy to fill in the *Evidence* and *Science knowledge* columns of the *Our Design Plan* page.

Our Design Plan

Name: Date:

Design Decisions	Supports	
	Evidence	Science Knowledge

Diagrams

Communicate Your Plan

Plan Briefing

You will present your design plan to your class in a *Plan Briefing*. A *Plan Briefing* is a little like a *Solution Briefing* and a little like an *Investigation Expo*. You will present a plan for a solution. As in an *Investigation Expo*, you will use a poster to organize your presentation.

As in a *Solution Briefing*, a *Plan Briefing* gives you a chance to get advice and suggestions from others. Their advice might help you find a better solution than you could have done by just using the ideas of your group.

You will get good advice from people if they understand why you made each of your decisions. The design plan you recorded on your *Our Design Plan* page should help you prepare your poster for this *Plan Briefing*.

Be a Scientist

Introducing a *Plan Briefing*

Preparing a *Plan Briefing* Poster

A *Plan Briefing* is much like the other presentations you learned to do. In a *Plan Briefing*, you present your design plan. You must present it well enough so your classmates can understand your ideas. They should be able to identify if you have made any mistakes in your reasoning. Then they can provide you with advice before you begin constructing your

△ Guide

When students finished their design plans, introduce the class to *Plan Briefings*. These are similar to *Investigation Expos*, but in *Plan Briefings* groups present details of their design plans.

solution. As a presenter, you will learn the most from a *Plan Briefing* if you can be specific about your design plans and about why you made your design decisions. You will probably want to draw pictures, maybe providing several views. You want everyone to know why you expect your design to achieve the challenge.

The following guidelines will help you as you decide what to present on your poster:

- Your poster should have a detailed drawing with at least one view of your design. You might consider drawing multiple views so the audience can see your design from different angles. It is important that the audience can picture what you are planning to build.
- Parts of the design and any special features should all be labeled. The labels should describe how and why you made each of your design decisions. Show the explanations and recommendations that support your decisions. Convincing others that your design choices are quality ones will convince them that you are making informed decisions backed by scientific evidence.
- Make sure to give credit to groups or students who ran the investigations you used in your design or who gave you ideas that helped your design.
- If another group provided an explanation or evidence that you are using, you should credit them with their assistance in developing your design.

Participating in a *Plan Briefing*

A *Plan Briefing* is similar to an *Investigation Expo* and a *Solution Briefing*. However, this time you will be presenting your design plan. As in other presentation activities, groups will take turns making presentations. After each presentation, the presenting group will take comments and answer questions from the class.

When presenting, be very specific about your design plan and what evidence helped you make your design decisions.

Your presentation should answer the following questions:

- What are the important features of the design?
- What criterion of the challenge will it achieve? What makes your plan the right way to achieve that criterion?

META NOTES

Assess students' skills in sharing ideas, asking questions, and responding to peers. Look for whether students base their reasoning about design choices on what they have learned about how different erosion-control methods work in models of the basketball-court site.

META NOTES

Keep track of revisions groups should make when they revise their plans so that you may check for them later.

TEACHER TALK

"Before you build your basketball-court solution, you're going to share your plans with the rest of the class. This way, your classmates may see things you missed in your plans, or you might realize as you're explaining something that it won't work. Then you can fix your plans before you build your basketball-court solution. To show your plans to the class, you're going to make a *Plan-Briefing* poster."

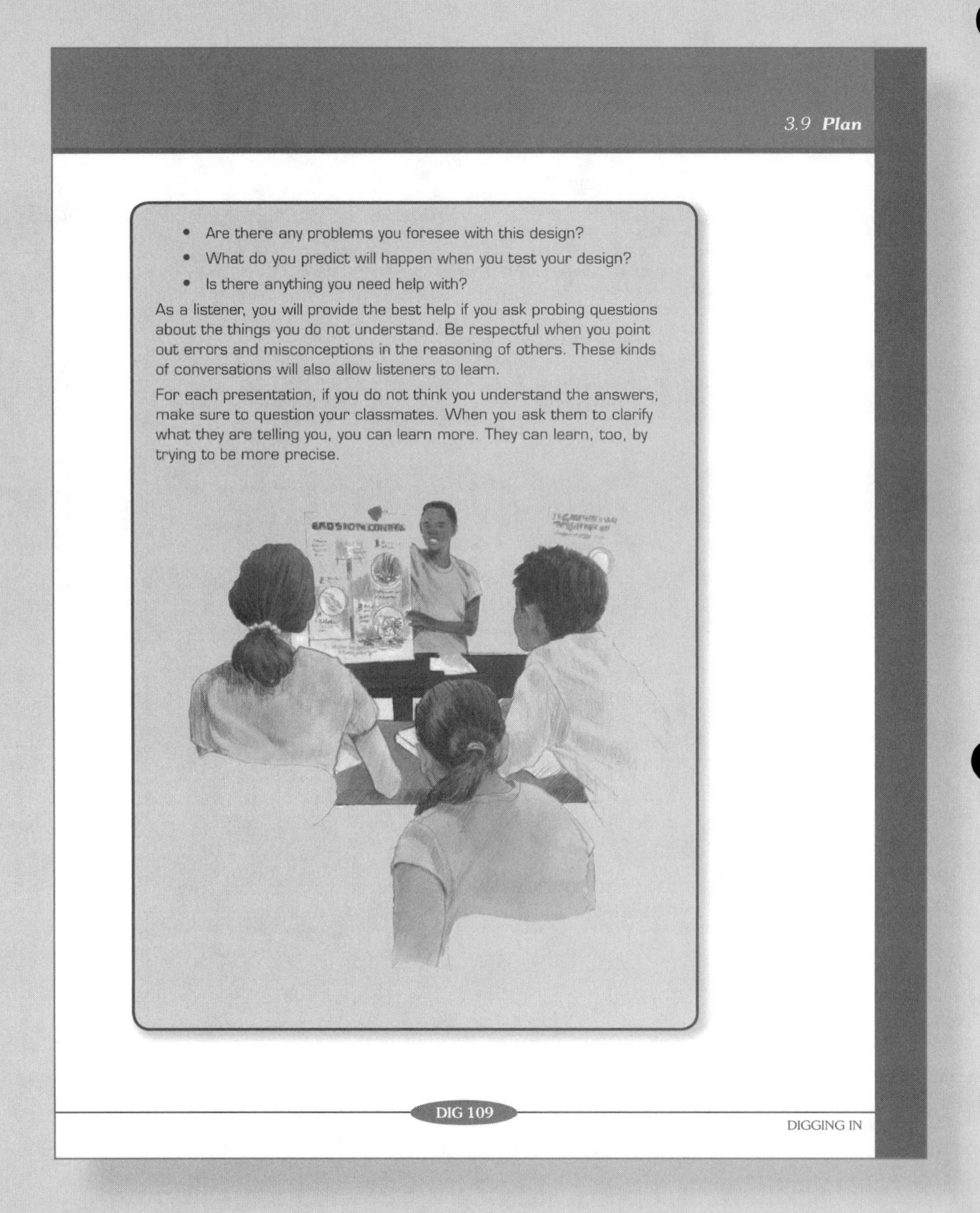

- Are there any problems you foresee with this design?
- What do you predict will happen when you test your design?
- Is there anything you need help with?

As a listener, you will provide the best help if you ask probing questions about the things you do not understand. Be respectful when you point out errors and misconceptions in the reasoning of others. These kinds of conversations will also allow listeners to learn.

For each presentation, if you do not think you understand the answers, make sure to question your classmates. When you ask them to clarify what they are telling you, you can learn more. They can learn, too, by trying to be more precise.

Tell students they will make posters with detailed drawings of their plans with all parts labeled. The class will discuss what criteria the features of the design achieve, what is expected to happen, and any possible problems. Emphasize that *Plan Briefings* should be specific and contain the reasoning behind design choices. They should also give credit where the results of an experiment or a group's recommendations are used. Students can use the guide in the *Be a Scientist* textbox as they prepare for and participate in *Plan Briefings.*

⬡ Get Going

Distribute posters for the *Plan Briefing*, give students a time frame (about 20 minutes), and get them started making their posters.

◇ Evaluate

As groups work on their *Plan Briefing* posters, look at groups' posters to see if their drawings are clear and if there is enough detail that the class will be able to discuss what criteria the plan will achieve and possible problems.

⬡ Get Going

When groups have finished their posters, have each group present their design plan to the class.

△ Guide

Tell the class that as each group presents, they should ask probing questions and politely point out errors and misconceptions. Emphasize that students should look for how the features of the design will achieve the criteria of the *Basketball-Court Challenge* and look for any problems that might come up. Model the participation you expect by asking questions of the presenting group when anything is not clear. Emphasize that the presenting group can ask the class for advice.

After each presentation, you may need to ask a question or two to begin the discussion. Ask a student to ask a question. These questions should be about how features of the design will achieve criteria of the *Basketball-Court Challenge*, what will probably happen when the design is implemented, and any problems that the group presenting may not have thought about. If the discussion stalls or loses focus, you can ask students whether all of the questions in the student text have been answered. Make sure the presenters respond to the students who asks the questions.

Revise Your Plan

10 min.

Groups revise their plans based on feedback from their Plan Briefing.

Revise Your Plan

You may have received some good advice from classmates about how to make your design plan better. If so, spend some time with your group doing that. Revise your *Our Design Plan* pages to match your revised plan. Add any evidence and science knowledge that supports your design decisions.

What's the Point?

People have strong opinions about some things. They often assume that their points of view are obvious to others. They think that other people will automatically hold those same views. But often they are surprised. What is obvious to you might look very different to someone else. You probably struggle with this all the time. What seems to work is to describe precisely why you believe what you believe each time you express an opinion.

Evidence that supports your point of view helps other people see your point. When people state their opinions without presenting evidence that justifies them, others are more likely to question their viewpoints. Whenever you need to convince someone of something, or when you are trying to decide between several alternatives, presenting evidence that supports a point of view is critical.

The same is true in convincing yourself that you have made a good decision. You should be able to justify a decision with evidence. Then you will be sure of your decision and more likely to make good decisions.

When you are planning the design of a product or process, it is often useful to hear from others. They can help you see how well your design meets the criteria of the challenge. If you present reasons for the decisions you are making, others can help you identify misconceptions you might have. They can also help you be more confident about your decisions. They can help you judge your decisions based on evidence and knowledge.

An important benefit of a *Plan Briefing* is that teams can learn from each other. A team may have struggled with one aspect of its design. That team may now have good advice for those who have not yet tackled that problem. They, in turn, may benefit from experiences some other team had.

Get Going

Now that groups have presented their designs and received feedback from the class, they can use the feedback to revise their plans. Give students a time frame and have them revise their plans. Emphasize that they need to record their new plans and justify their revisions.

☐ Assess

As groups are revising their plans, check to see what revisions groups are making. They should be using the feedback from the class.

Assessment Options

Targeted Concepts, Skills, and Nature of Science	How do I know if students got it?
Scientists often work together and then share their findings. Sharing findings makes new information available and helps scientists refine their ideas and build on others' ideas. When another person's or group's idea is used, credit needs to be given.	**ASK:** What are some ways you can discover flaws in your design before you begin building? **LISTEN:** Students should include discussion with peers among their responses.
Scientists must keep clear, accurate, and descriptive records of what they do so they can share their work with others and consider what they did, why they did it, and what they want to do next.	**ASK:** How did you learn from hearing the reasons for other groups' decisions? **LISTEN:** Students should have learned which design choices were more effective than others and why.
Erosion is the process of soil and other particles being displaced by water, waves, wind, and gravity.	**ASK:** What kinds of erosion does you solution address? **LISTEN:** Students' solutions should address erosion that is likely to be found at the site of the basketball court (most likely, heavy rain).

Teacher Reflection Questions

- What types of problems did groups learn about during their *Plan Briefing?*
- How did you model the kinds of questions and comments you expect from students during presentations? What ideas do you have to encourage appropriate discussion among students?
- How was managing the *Plan Briefing* different from managing presentations of recommendations and explanations? Is there anything you would do differently next time?

NOTES

SECTION 3.10 INTRODUCTION

3.10 Build and Test

Build and Test Your Basketball-Court Solution

◀ *$1\frac{1}{2}$ class periods*

A class period is considered to be one 40 to 50 minute class.

Overview

Students build their basketball-court design solutions, test them, and revise them based on the results of their tests. Using the same materials they used for their simulations, students construct a model of the land, the basketball court, and their erosion control method(s) based on their design plans. They test their design and revise the solutions, and continue testing and revising as necessary, using the same procedure to test each time. They share their designs with the class, as well as the different design ideas they tried and why they chose the one they did.

Targeted Concepts, Skills, and Nature of Science	Performance Expectations
Scientists often work together and then share their findings. Sharing findings makes new information available and helps scientists refine their ideas and build on others' ideas. When another person's or group's idea is used, credit needs to be given.	Students should work with their groups to build and revise their solutions and share them with the class.
Scientists must keep clear, accurate, and descriptive records of what they do so they can share their work with others and consider what they did, why they did it, and what they want to do next.	Students should record the revisions they make to their designs.
Erosion is the process of soil and other particles being displaced by water, waves, wind, and gravity.	Students should revise their solutions as necessary to keep the soil at the basketball-court site from being eroded.

Materials	
1 per group	Stream table tray Small cup Spray bottle, filled with water Ruler Plastic drop cloth Drain bucket
1 bin per group	Native soil Spanish moss Slate chips Landscape materials
1 per classroom	Paper towel roll
3 per student	*Running Our Erosion-Control Model* page
1 per student	Disposable gloves *Our Simulation* page
1 per class (optional)	Digital camera

Activity Setup and Preparation

Consider bringing digital cameras to class so that students can photograph the design ideas they test.

Before class begins, organize materials and equipment to allow students to expeditiously run their investigations. Prepare containers of soil to distribute to groups. Students may need to drain water and rebuild the slope after each test.

Homework Options

Reflection

- **Science Content:** Did any groups' final designs surprise you? What were some of the differences between your design and your classmates' designs? Which do you think would be most effective at the actual basketball-court site? *(Students should recognize any major differences between their final designs and other groups' final designs. They should evaluate which designs would probably work best based on what they know about erosion.)*

SECTION 3.10 IMPLEMENTATION

◀ $1\frac{1}{2}$ *class periods**

3.10 Build and Test

Build and Test Your Basketball-Court Solution

You planned your best design based on evidence you have available. You presented it to others. You received advice from your classmates. You might have revised your plan based on what your classmates suggested. You are now ready to test your erosion-control solution. You will test it by building a model and then simulating rainfall on the model to see how well it prevents the hill from eroding. You hope to have the erosion-control solution that works the best. If you do, the school board will accept the donated land, the basketball court will be built, and your solution will be implemented!

Build Your Basketball-Court Solution

Work with your group to model and test your solution. You will have the opportunity to revise and test your model two or three times. After you complete your second or third iteration, the class will hold a demonstration and competition. Each group will demonstrate their basketball court solution in front of the class. Each group will also present to the class the changes they made in their design since the *Plan Briefing*. They will explain why they made those changes. When you recommend an erosion-control solution to the school board, you will need to tell them not only how to design the erosion-control solution, but also why you think that design is the best one. On the next pages are some hints for you about how to manage iteration to design your best solution for the basketball court.

3.10 Build and Test

Build and Test Your Basketball-Court Solution

5 min.

Students now have a chance to test their designs for the Basketball-Court Challenge *solution.*

○ Engage

Motivate students to build their best design and let them know they may be revising their designs.

*A class period is considered to be one 40 to 50 minute class.

TEACHER TALK

"You've got a design plan, now you will be able to build and test it and you'll be able to revise it too. Remember that your goal is to have the erosion-control solution that works best so that the school board will accept the donated land and build a basketball court on it."

Build Your Basketball-Court Solution

15 min.

Students build and test their basketball-court solution.

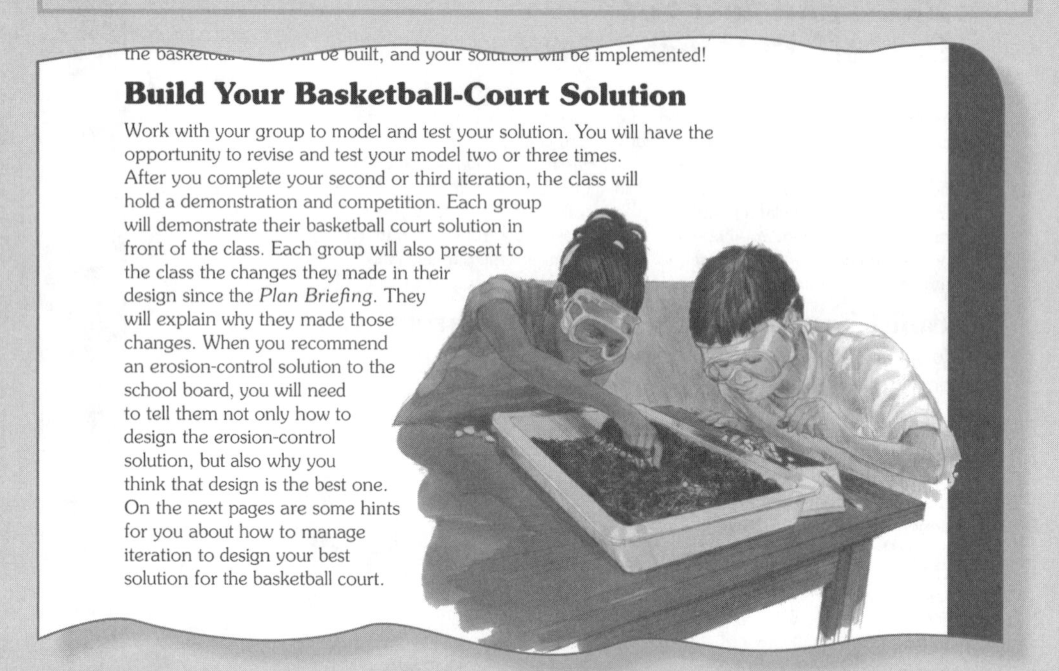

the basketball ... will be built, and your solution will be implemented!

Build Your Basketball-Court Solution

Work with your group to model and test your solution. You will have the opportunity to revise and test your model two or three times. After you complete your second or third iteration, the class will hold a demonstration and competition. Each group will demonstrate their basketball court solution in front of the class. Each group will also present to the class the changes they made in their design since the *Plan Briefing*. They will explain why they made those changes. When you recommend an erosion-control solution to the school board, you will need to tell them not only how to design the erosion-control solution, but also why you think that design is the best one. On the next pages are some hints for you about how to manage iteration to design your best solution for the basketball court.

△ Guide

Let students know they will now build and test their basketball-court solutions. Emphasize that they should not begin testing their design until after you have given them approval to test it.

⬡ Get Going

Let students know how long they have to build their designs, and then get students started.

Have students test the integrity of their models by saturating the soil with water and then simulating light and heavy rain. Their model should show a realistic level of erosion. If they have a landslide after a light misting, or if they have no erosion even after dumping water on the model, they will need to adjust it.

◇ Evaluate

Before groups test their design, visit each group and make sure their model was approved. Test the integrity of the model by simulating soft rain (spraying from a spray bottle on the model). Make sure you discuss the next segment with students before they begin testing.

Test Your Basketball-Court Solution

Below are some suggestions for testing your designs. Use *Testing My Design* pages to record your work.

Testing Your Designs

In experiments, it is important to run several trials. Then you can be sure your results are consistent. The same is true in testing a design. Each time you test a design, make sure to run enough trials. Choose the number of trials that will allow you to see how it performs. Follow the same procedure each time you test it. Otherwise, you will not know if the design is causing the effects you see or if something you did not control in your procedure is responsible for your results.

Recording Your Work

As you test and revise your design, it will be important to record the results of your tests. You will also need to record the changes you are making. You should record why you are making those changes. This is for several reasons:

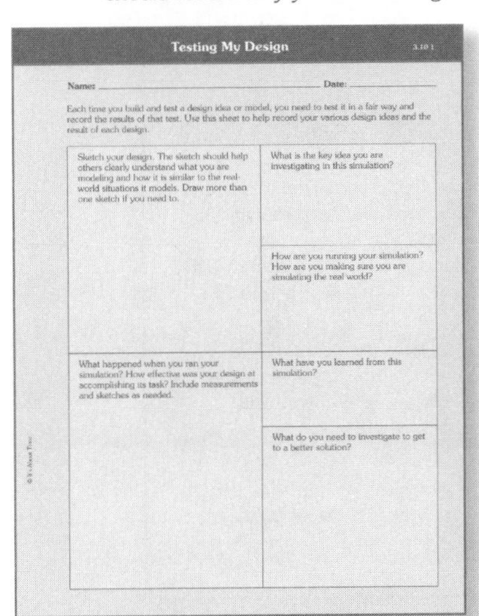

Testing My Design 3.10 1

Name: ______________________ Date: ____________

Each time you build and test a design idea or model, you need to test it in a fair way and record the results of that test. Use this sheet to help record your various design ideas and the result of each design.

Sketch your design. The sketch should help others clearly understand what you are modeling and how it is similar to the real-world situations it models. Draw more than one sketch if you need to.	What is the key idea you are investigating in this simulation?
	How are you running your simulation? How are you making sure you are simulating the real world?
What happened when you ran your simulation? How effective was your design at accomplishing its task? Include measurements and sketches as needed.	What have you learned from this simulation?
	What do you need to investigate to get to a better solution?

- Sometimes, what seems like a mistaken approach turns out to work better when some other part of the design is changed.
- You may need to remember what you did and did not test.
- You can use your earlier designs to help teach others.
- By studying your earlier designs, you can learn how your mistakes and successes contributed to your science understanding.

Use *Testing My Design* pages to record your results, changes, and the reasons for changes. Use one page for each iteration.

You can do as many iterations of your erosion-control design as you have time for. However, remember that it is important to run each trial

△ Guide

After you have given groups approval to test their designs, briefly discuss the importance of testing procedures with the group. Emphasize that it will be important to test each design the same way they did in the previous design. It is important to run multiple trials for each design. Each group should decide on a procedure at the beginning, specifying the number of trials, how to run the simulation, and how to measure the results.

Test Your Basketball-Court Solution

30 min.

Groups iteratively test and revise their basketball-court solutions.

META NOTES

Provide guidance to each group individually rather than to the entire class as they begin this segment.

META NOTES

It is important that groups use the same procedure each time they run their test. If you see groups changing their procedure, remind the class that in addition to running multiple trials, they need to make sure that each test and each trial is run the same way.

META NOTES

As students revise their plans, remind them it is more effective to change one small thing in the design for each iteration than to make many changes at once. This allows students to isolate specific things that work or do not work.

the same way for every design. For example, you must pour the same amount of water onto the model in the same way for every trial. And remember that you need to pour the water to simulate real-world rainfall as well as possible.

Be a Scientist

Iteration

Remember that iteration is a process of making something better over time. That something may be a product, a process, or an understanding. Scientists and student scientists iteratively understand new concepts better over time. Scientists iteratively make investigative procedures better over time (as you did in the *Lava Flow* activity).

Designers iteratively make designs better over time. Each time they test a design, they might find ways to improve it. That is what you are doing now. Sometimes a design does not work as well as the designer expected. When that happens, the designer tries to understand why it is not working as well as expected and makes changes based on that analysis. If your design doesn't work as well as you wanted it to, your first feeling may be to throw away those failed plans and begin again. Do not! If you began with a design based on evidence and science, then your solution will probably work well with some changes. Designers only throw away designs and begin again if the problems are so big that it would be easier or less expensive to begin again.

You saw the power of iteration earlier in this Unit. In the *Build a Boat Challenge*, you improved the design of your boat. You built the boat and tested it. You identified weaknesses and improved your design. Now, with your erosion-control solution, you will again have the opportunity to iteratively enhance your design.

Usually, the best way to iterate is to make one revision at a time. If you make and test one change at a time, then you will know the effect of that change.

It is important to record the results of each test and all the changes made. Each group should record their results for every iteration on a *Testing My Design* page.

As students are working on their designs, you may choose to have them briefly present to the class. Showing the class the designs they are working on and how they have modified their plans allows the rest of the class to give or receive advice.

As groups are testing their designs, you can ask what ideas they have tried out. If any groups seem stuck on a design and do not think they can improve it, you might ask them what they know about alternative designs. Would removing a feature of their design make it work better or worse? Would adjusting a feature make it work better?

NOTES

Communicate Your Design: *Solution Briefing*

20 min.

Students present in a Solution Briefing *to the class.*

Communicate Your Design

Solution Briefing

While you are working on iterating toward a better design, your teacher might have you present your design-in-progress to the class in a short *Solution Briefing*. You will also have a *Solution Briefing* when the time for testing your solutions is over. Recall that a *Solution Briefing* is very much like a *Plan Briefing*. You present your solution for others to comment on.

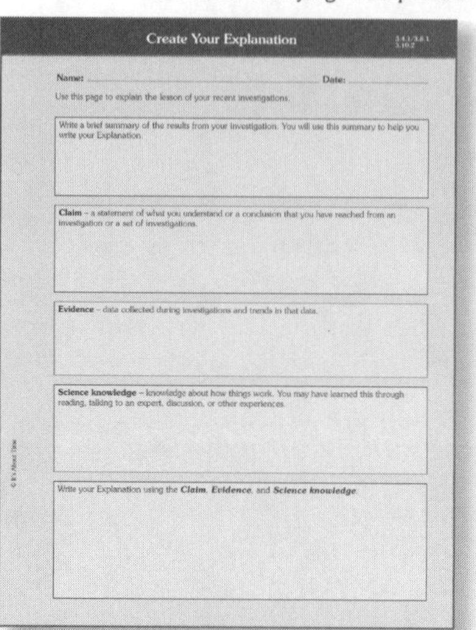

Create Your Explanation

Name: ______________________ Date: __________

Use this page to explain the lesson of your recent investigations.

Write a brief summary of the results from your investigation. You will use this summary to help you write your Explanation.

Claim – a statement of what you understand or a conclusion that you have reached from an investigation or a set of investigations.

Evidence – data collected during investigations and trends in that data.

Science knowledge – knowledge about how things work. You may have learned this through reading, talking to an expert, discussion, or other experiences.

Write your Explanation using the ***Claim***, ***Evidence***, and ***Science knowledge***.

An important issue you will have to pay attention to when you engage in these *Solution Briefings* is time. Because you have to break down your designs at the end of each class period, each presentation will have to be short to fit them all in. When your teacher calls a *Solution Briefing,* be prepared to quickly present your progress. Describe the design you are working on. Tell the class how it is different from what you thought you were going to build. Tell them why it is different. Show them what happens when you pour rain on it. Tell them anything you are having trouble with, and ask for advice. Your group's experience may provide valuable lessons for others. If you are having trouble, a *Solution Briefing* is a chance to get help.

Remember, you can learn from attempts that did not work as well as you expected. So do not be shy about presenting what has not worked as well as you expected. You and others can learn from mistakes. Your peers can give you advice about design, construction, and testing.

Reflect

Discuss your erosion-control solution and what you learned from the *Solution Briefing*.

△ Guide

Once students have finished iteratively building and testing their designs, let them know they will briefly present their design to the class.

Remind students of the information a *Solution Briefing* should convey: the original design plan, the history of the group's revisions, the way the group used the recommendations of the class, and the final design. Emphasize that it should also detail the changes the group made in each iteration and the reasons for the changes. Let students know how long they will have for their *Solution Briefing* and give them a few minutes to prepare their presentations.

Have each group briefly present their design solution. As each group presents, encourage students to ask questions and model the kinds of questions students should be asking. These should be about what techniques groups tried, what criteria the designs achieve, how the constraints were or were not accounted for in the design, and whether there are any problems with the design.

⬡ Get Going

Give groups a few minutes to answer the *Reflect* questions on their own. Lead a class discussion of their answers. Listen for the following:

1. Students should evaluate how well their solution meets each of the criteria and constraints they identified. These should include:

   Criteria: must prevent erosion at the top of the hill from covering a 28 m X 15 m basketball court at the bottom of the hill.

   Constraints: the basketball court must be 5 m from the base of the hill; project must not damage houses (30 m X 10 m) that are 12 m from the court on either side; height of the hill is 10 m.

2. Students should have documented the changes they made in their *Testing My Design* pages.
3. Students should recognize ways they used ideas that came from class discussions.
4. Students should honestly evaluate the ways in which their solutions need more work.
5. Students should have ideas for new recommendations. Use these to guide students to begin thinking about how to update the *Project Board*.

Reflect

15 min.

Students answer the Reflect *questions and discuss their answers with the class.*

Update the *Project Board*

5 min.

The class updates their Project Board.

Answer the following questions and be prepared to discuss your answers with the class:

1. How well does your design meet the criteria and constraints of the *Basketball-Court Challenge?*
2. What changes did you make to your design to improve its ability to control erosion?
3. Describe any ideas you got from other groups' designs and presentations or recommendations you used to improve your own design.
4. What changes do you think you still need to make to your design to be more successful?
5. You probably can make some new recommendations about managing erosion, this time about combining erosion-control methods with each other. What new recommendations, if any, should be added to the *Project Board*? Develop your recommendations and their supporting explanations using *Create Your Explanation* pages.

Update the *Project Board*

Based on your experiences combining erosion-control methods with each other, you have derived some new recommendations. Update the *Project Board* with any new recommendations you have and any new evidence you have collected that can help support them.

What's the Point?

You have just built and tested your erosion-control solution. You have run several simulations of your solution and followed the same procedure each time. Each time, you recorded your work and revised your design if it did not work as you expected. You also recorded why you made changes to your design before each new iteration. You presented your design-in-progress to the class in a *Solution Briefing*, asked for advice, and listened to the presentations of others. Based on your experiences combining erosion-control methods, you developed new recommendations and supporting explanations and added them to the *Project Board*. Modeling and simulation is used to test solutions when it would be too dangerous or too expensive to test solutions in the real world. It is always important to be able to predict how a solution will perform before building it. It is also possible to learn how to make a solution better by modeling it.

△ Guide

Ask the class what new recommendations they have to add to the class *Project Board*, and update the *Project Board* as students give you information.

Assessment Options

Targeted Concepts, Skills, and Nature of Science	How do I know if students got it?
Scientists often work together and then share their findings. Sharing findings makes new information available and helps scientists refine their ideas and build on others' ideas. When another person's or group's idea is used, credit needs to be given.	**ASK:** How did you use things you learned from other groups as you revised your solution? **LISTEN:** Students should have used things they learned from other students' simulations to guide their design decisions.
Scientists must keep clear, accurate, and descriptive records of what they do so they can share their work with others and consider what they did, why they did it, and what they want to do next.	**ASK:** What parts of your investigation should you record? **LISTEN:** Students should indicate that the results and the changes in procedure are important to record.
Erosion is the process of soil and other particles being displaced by water, waves, wind, and gravity.	**ASK:** What effects of erosion were hardest to control? **LISTEN:** Groups should use their observations to describe effect of erosion they had trouble controlling.
Scientists use models to simulate processes that happen too fast, too slow, on a scale that cannot be observed directly (either too small or too large), or that are too dangerous.	**ASK:** What questions were you able to answer using a simulation that you would not otherwise have been able to answer? **LISTEN:** Students should recognize they would not have been able to test which combination of erosion-control methods would be effective without the help of models.

Teacher Reflection Questions

- Did students apply lessons from the previous *Learning Sets* to their iteration? Did they make small changes for each iteration?
- How did you assess how students were giving and receiving feedback and building on each other's ideas? What indicators could you look for next time?

ADDRESS THE BIG CHALLENGE INTRODUCTION

1 class period ▶

A class period is considered to be one 40 to 50 minute class.

Address the *Big Challenge*

Advise the School Board

Overview

Students revise their plans and share their revised plans with the class using a *Plan Showcase*. The class discusses the strengths and weaknesses of each design and decides what the class's final recommendation should be. The class updates the *Project Board*, recording their final recommendations and the things they learned that support those recommendations. Finally, they write a letter to the school board presenting their recommendations.

Materials	
1 per student	*Our Design Plan* page
1 per class	Class *Project Board*

Homework Options

Reflection

- **Science Process:** You are on a team of engineers designing a moon probe that will sit in a lunar crater and record data. The probe will be powered by solar panels. Someone has suggested that the solar panels might slowly be covered over by sand from the wind. How would you research this possibility? *(Students might suggest to investigate how past lunar probes have been affected by wind and sand and what kinds of observations humans have made on lunar walks. In other words, collecting cases and case studies. Based on what they learned, they could design a model and simulate what would happen to the solar panels of the lunar probe on the surface of the moon.)*

ADDRESS THE BIG CHALLENGE IMPLEMENTATION

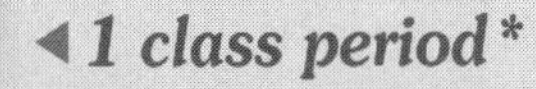

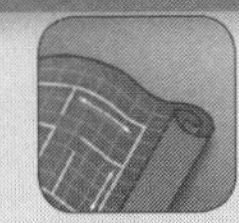

Address the Big Challenge

Advise the School Board

Your Challenge for this *Learning Set* was to come up with a solution to the erosion problem surrounding a proposed basketball court. You need to make a recommendation to the school board showing how erosion can be controlled without affecting the nearby houses. Your class might want to send one proposed solution to the school board, or you might want to send a set of proposed solutions. You might want to send your set of recommendations. What is important is that you send the school board a package that will convince them to accept the donated land and build the basketball court.

Each group will begin by revising their design one last time based on suggestions made earlier and then presenting their recommendations to the class in a *Plan Showcase*. The class will then decide what solution or set of solutions to send to the school board. The class will work together to produce a good package.

Address the Challenge

You have had experience modeling and simulating your proposed solution to the *Basketball-Court Challenge*. Some of what you proposed earlier worked well, and some of what you proposed did not work as well as you thought it would. It is time now to revise your design one last time and make your recommendation to the school board.

Use an *Our Design Plan* page to record your design and your decisions. Draw as many diagrams of your design as you need. Record your design decisions and the evidence and science knowledge that support them in the appropriate columns. Use the class's *Project Board* to help support your recommendations. The *Project Board* has on it the recommendations your class developed and the evidence and science knowledge used to create those recommendations.

Address the Big Challenge

Advise the School Board

< 5 min.

○ Engage

Let students know that they will be revising their recommendations and sharing them with the class before writing to the school board.

*A class period is considered to be one 40 to 50 minute class.

TEACHER TALK

"It's important that before we decide to send anything to the school board, we make sure we have picked out the best possible recommendation and have included all the supporting evidence and science knowledge to back it up. You heard different recommendations during the class's *Solution Briefing*. Now you will have a chance to revise your recommendation and design plan, then we'll share them with the class and finally come up with a class recommendation."

Address the Challenge

10 min.

Students revise their plans.

set of solutions to send to the school board. The class will work together to produce a good package.

Address the Challenge

You have had experience modeling and simulating your proposed solution to the *Basketball-Court Challenge*. Some of what you proposed earlier worked well, and some of what you proposed did not work as well as you thought it would. It is time now to revise your design one last time and make your recommendation to the school board.

Use an *Our Design Plan* page to record your design and your decisions. Draw as many diagrams of your design as you need. Record your design decisions and the evidence and science knowledge that support them in the appropriate columns. Use the class's *Project Board* to help support your recommendations. The *Project Board* has on it the recommendations your class developed and the evidence and science knowledge used to create those recommendations.

⬡ Get Going

Have groups revise their plans using *Our Design Plan* pages. Emphasize that they should use the recommendations, evidence, and science knowledge on the class *Project Board* to support their final recommendations.

NOTES

Communicate Your Design

Plan Showcase

Later, you will write a letter to the school board. For now, prepare to present your solution and recommendations to the class in a *Plan Showcase*. A showcase is a presentation that shows off your solution to a challenge or answer to a question. Make a poster that includes the design you recommend and the reasoning behind the decisions you made—the evidence and science knowledge that support them.

You will share your final design in a *Plan Showcase*. Be prepared to share with your class your solution, the reasoning behind it, why you think it is a good solution (even if it does not work 100%), and how you came to your solution. Think of your presentation as practice for presenting to the school board. Remember that your reasoning will be important to the school board as they make their final decision as to whether or not they should accept the donated land and build the basketball court.

Be a Scientist

Introducing a *Showcase*

A *Showcase* is for the purpose of presenting your solution to a challenge or problem or your answer to a question. In a *Plan Showcase*, you showcase your ideas about how to achieve a challenge. You will hold *Solution Showcases* in other Units where you will present actual built solutions. Whichever kind of *Showcase* you participate in, the purpose is to present your solution or answer and help your audience understand what makes it a good solution or answer. In general, you present five things in a *Showcase*:

- the challenge you were addressing or question you were answering, including criteria and constraints;
- your solution;
- the reasoning behind your solution (evidence and science knowledge that support your decisions);
- an analysis of how well your solution addresses the criteria and constraints or answers the question (including what issues you did not address well yet); and

Communicate Your Design: *Plan Showcase*

15 min.

Groups present their final designs using Plan Showcases.

△ Guide

Once students have finished their final designs, let them know they will present their final designs to the class in a *Plan Showcase*.

Introduce a *Plan Showcase*, highlighting the information it should convey: the original design plan, the history of the group's revisions, the way the group used the recommendations of the class, and the final design. Emphasize that it should detail the changes the group made in each iteration and the reasons for the changes. Let students know how long they will have

- the history of how you got to your solution, including, for each iteration, the testing you did, your results, your explanations of results, and how that led to your next iteration.

As you listen, it will be important to look at each design carefully. You should ask questions about how the design meets the criteria of the challenge. Be prepared to ask (and answer) questions such as these:

- How well does the design meet the goals of the challenge?
- How did the challenge constraints affect the use or success of this design?
- What problems remain?

Reflect

You will discuss the suggested plans as a class and decide what you should present to the school board. Work in your small group first to identify the strengths and weaknesses of each design. Be prepared to discuss your answers with the class.

- **Which design addresses the challenge the best? Why?**
- **If there is more than one that addresses the challenge well, compare them to each other. What are the strengths and weaknesses of each?**

Reflect

10 min.

Discuss the strengths and weaknesses of the designs with the class.

for their *Plan Showcases*, and give them a few minutes to prepare their presentations. Point out that there are guidelines they can use in the student text.

○ Get Going

Have each group briefly present their *Plan Showcase.* As each group presents, encourage students to ask questions and model the kinds of questions students should be asking. These should be about what techniques groups tried, what criteria the designs achieve, how the

- Given the strengths and weaknesses of each design, what do you think should be included in the package to the school board?

Use evidence to support your reasoning.

Update the *Project Board*

Now that you have completed the *Basketball-Court Challenge* and decided what to send to the school board, it is time to go back to the *Project Board* for one final update. You will focus mainly on the middle and last columns, filling in what you have learned about the causes of erosion and methods that work well to manage it. Add recommendations to the last column, based on your discussions about solutions and what seemed to work well in controlling erosion at the basketball site.

Advise the School Board

Write a report with your recommendations to the school superintendent, letting her know your best recommendations about the *Basketball-Court Challenge*. Remember that good advice includes recommendations supported with evidence and science knowledge. Your reasoning will be important to the school board as they make their final decision as to whether or not they should accept the donated land and build the basketball court.

Update the *Project Board*

5 min.

Update the Project Board *with what students have learned and their recommendations.*

Advise the School Board

10 min.

Students write letters with recommendations to the school board.

META NOTES

You can also have students write the letter together with their groups or with the class, or you can assign the letter for homework.

constraints were or were not accounted for in the design, and whether there are any problems with the design.

Some of the questions they might ask are at the bottom of *Introducing a Showcase.*

⬡ Get Going

Have groups meet to briefly discuss the strengths and weaknesses of each design. Emphasize that they should use the bulleted questions in the student text as guidelines and they should support their ideas with evidence.

△ Guide

Initiate a class discussion of what the final recommendation to the school board should be. As groups present their ideas, emphasize that they should use evidence to support their reasoning.

◇ Evaluate

By this time, the class should be coming to agreement about what design features are effective.

△ Guide

When the class has decided on which recommendations to send to the school board, turn students' attention to the *Project Board*. Ask students what they have learned about the causes of erosion and ways to control it. Ask them what evidence they have. Update the third and fourth columns of the *Project Board* as students suggest ideas. Finally, ask them what recommendations they should put in the last column of the *Project Board*.

⬡ Get Going

Have students write letters with recommendations to the school board. Emphasize that the recommendations should be clear and specific, supported by evidence and science knowledge.

◇ Evaluate

Students' letters should demonstrate the reasoning that connects the recommendations to the evidence and science knowledge. They should also be specific enough that the erosion-control method could be constructed using the letter as a guide.

Teacher Reflection Questions

- What were the biggest changes that you observed in students' thinking about the challenge in the course of the *Basketball-Court Challenge?* Did you observe anything you did or the students did that made these changes possible?
- What evidence do you have that students used evidence from the *Project Board* to guide their decision-making? What can you do to help them make use of this?
- How did you manage the discussion of what recommendations to make to the school board? How would you, or did you, resolve disagreements at this stage?

ANSWER THE BIG QUESTION INTRODUCTION

Answer the *Big Question*

How Do Scientists Work Together to Solve Problems?

◀ ***1 class period***

A class period is considered to be one 40 to 50 minute class.

Overview

Students watch a video of real-life designers meeting a design challenge by researching the variables involved, identifying criteria and constraints, trying various ideas, and collaborating to arrive at a final design. Students compare what the designers do in the video to what they did in class. Discussing the *Stop and Think* questions and writing answers to the *Reflect* questions, they use what they learned through their projects and the video to answer the *Big Question* of the Unit: *How do scientists work together to solve problems?*

Materials	
1 per class	DVD player and television *IDEO Deep Dive* Video

Activity Setup and Preparation

Have a DVD player and television ready to show the video before class.

Preview the video before showing the class.

Homework Options

- **Science Process:** How would the design process have been different if you skipped the experiments at the beginning of the Unit? *(Students' answers should reflect a realistic assessment of how the results of the experiments were useful.)*

1 class period ▶

Answer the Big Question

How Do Scientists Work Together to Solve Problems?

5 min.

Students review the Unit question and reflect on how they have been working together to solve challenges.

META NOTES

If you remember a specific instance(s) of groups solving a problem by working together this would be a good example to discuss with the class.

ANSWER THE BIG QUESTION IMPLEMENTATION

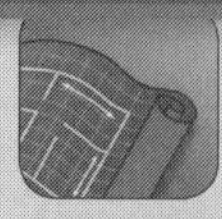

Answer the Big Question

How Do Scientists Work Together to Solve Problems?

You began this Unit with the question: *How do scientists work together to solve problems?* You addressed several small challenges. As you worked on those challenges, you learned about how scientists solve problems. You will now watch a video about real-life designers. You will see the people in the video engaging in activities very much like what you have been doing. You will then think about all the different activities and reasoning you have done during this Unit. Lastly, you will write about what you have learned about doing science and being a scientist.

Top: A trio of IDEO designers reviews a proposed concept framework together. ***Middle:*** *A project team compares a series of models for a skate park layout.* ***Bottom:*** *The informal atmosphere of a lounge area acts as a backdrop to a group brainstorm.*

Watch

IDEO Video

The video you will watch follows a group of designers at *IDEO. IDEO* is an innovation and design company. In the video, *IDEO* designers face the challenge of designing and building a new kind of shopping cart. These designers are doing many of the same things that you did. They also use other practices that you did not use. As you watch the video, record the interesting things you see.

Project-Based Inquiry Science DIG 120

○ Engage

Initiate a discussion of what the class has learned about working together to solve problems. Students might say they have learned to share results, to work in small groups to develop ideas or get data and then to get feedback, and to use standard procedures. Connect this with any *Solutions Briefings* you conducted during the *Basketball-Court Challenge* since they will be students' most recent experiences of presenting to get advice from the class. Ask what some real-world applications of these problem-solving strategies might be.

*A class period is considered to be one 40 to 50 minute class.

TEACHER TALK

"What are some of the problems you solved to control erosion? How about in the boat challenge? What are some of the things the class did that helped make it easier to solve those problems? What are some real-world challenges where you could use these skills?"

⬡ Get Going

Have students write letters with recommendations to the school board. Emphasize that the recommendations should be clear and specific, supported by evidence and science knowledge.

Let students know they will watch a video about a design team that solves these kinds of problems every day. In this video the team is working on the design of a shopping cart. Go over the *Stop and Think* questions with the class before they watch the video so they know what to look for in the video. Then show the video to the class.

The video lasts just over 20 minutes. Students should note information relevant to the *Stop and Think* questions while watching the video.

Watch

25 min.

The class watches the IDEO *video.*

NOTES

Stop and Think

15 min.

Students answer the Stop and Think *questions after watching the* IDEO *video.*

After watching the video, answer the questions on the next page. You might want to look at them before you watch the video. Answering these questions should help you answer the *Big Question* of this Unit: *How do scientists work together to solve problems?*

Stop and Think

1. List the criteria and constraints that the design team agreed upon. Which criteria and constraints did the team meet? In your opinion, what other criteria and constraints were not included in the team's discussion?
2. Why did the team split into smaller groups? What did the team hope to accomplish by doing this?
3. What types of investigations did you see the teams doing? What information were the teams trying to collect? Discuss how the information they collected helped the team design a better shopping cart.
4. Why do you think team members' ideas were not being criticized during the initial stages of design?
5. Give at least three examples from the video of how this group of people kept themselves on track to reach their goal on time. (How did they keep the project moving along?)
6. Analyze the team's final product. List three advantages and three disadvantages you see in the new shopping cart.
7. Compare the practices you saw in the video to the practices you used in the classroom. How are they different? How are they the same?
8. Give examples from the video of collaboration and design practices you did not use in the classroom.
9. List two aspects of the *IDEO* work environment you liked. List two aspects you did not like.

△ Guide

After the class has seen the video, have students write their answers to the *Stop and Think* questions and then meet with their groups to come up with the best answers.

Once groups have had time to write their answers, discuss them with the class. You can use the following answers to guide the discussion.

1. Students should be able to identify most of the criteria and constraints discussed in the video during the interview stage and the initial design discussion called *Dive In*. The criteria and constraints mentioned in the video are:

   Criteria: the design must be constructible, safe for children, discourage theft, must nest, make groceries easy to find, and it should not be likely to coast away in the wind.

   Constraint: the new shopping cart design must cost about the same as the traditional shopping cart.

   They should also be able to pick out the criteria and constraints met by the final design. The design team met the criteria for child safety, discouraging theft, nesting carts, easy to find groceries, not likely to move in the wind, and easy to move (wheels turned in all directions). They also met the constraint of costing the same as a traditional shopping cart.

   Students should come up with other criteria or constraints. These will most likely come from their own experiences with shopping carts. It is not mentioned that they should be durable in all weather conditions, or they should not require much cleaning, they should have high capacity.

2. Students should demonstrate they understand smaller groups can obtain background information faster and can focus on one aspect of the larger challenge. This keeps the designers focused and allows each group to optimize one idea and tends to be a more efficient use of time. The design team determined four focus areas to work on (safety, ease of checkout, ease of finding groceries in the cart, ease of shopping) and decided that a team should investigate and design a cart with just that feature in mind. They could pull the design features together into a single cart.

3. Students should note how in the first investigation designers were researching the problem. Students should describe how design teams investigated what problems shopping carts had and how people used them. They should mention the team members observed and interviewed shoppers using shopping carts, and they interviewed store employees who worked with the shopping carts about what kinds of issues they had. Students should include how they used the information obtained to plan designs addressing the problems they heard about.

4. Students may include how ideas that do not seem to work may become useful when combined with other ideas, how ideas that do not work may lead to ideas that do work, and how critiquing rather than criticizing ideas encourages open brainstorming.

5. Students' examples might include: the team broke into small groups that each focused one of the issues; team leaders occasionally orchestrated the efforts of the group; and the group voted on ideas to work with, narrowing its focus.

6. Students might point out some advantages are: it has no components with a lot of surface area (once the baskets are removed), it is unlikely to coast away in the wind, it allows you to leave the cart somewhere and take only the basket with you as you shop, it has a safety seat for children with a work area, the wheels move in all directions. Some disadvantages students might point out are: the plastic baskets may not be durable and may need to be replaced often, and the storage and cleaning of all the components may be tedious and time consuming, the cart may support paper bags to be hung from the hooks when leaving the store.

7. Students may notice the class and the design team did a lot of brainstorming; they both thought about what the criteria and constraints were, they both shared mockups after smaller groups had tackled individual design problems, and they both built on everyone's ideas to get the best solution. Students may describe differences of the *IDEO* starting with the whole group brainstorming general ideas, while the class only brainstormed as a group to was identify criteria and constraints and when groups presented their work to the class for feedback.

8. Student may notice the *IDEO's* team posted ideas on the walls and then voted on them.

9. Look for things students liked that might be useful principles. Look for things students did not like that might be things that could be changed or that were not essential to the design process.

10. Students may point out team members had to refrain from criticizing one another's ideas, team members had to stay focused and have one conversation at a time, and team members had to collaborate and build on each other's ideas.

11. Brainstorming in a classroom was most effective when students proposed ideas without fear of criticism. When groups focused on investigating one variable, it kept their conversations focused. Presentations the class gave the *Investigation Expos* and *Solution Showcases* allowed groups to build on one another's ideas.

PBIS *Answer the Big Question*

10. The IDEO workers have to take on extra responsibilities to maintain their fun, yet productive, work environment. Identify and discuss at least three of these responsibilities.
11. Relate the responsibilities you identified to working with a group in the classroom. Justify your choices using evidence.

Reflect

The following questions review the concepts you learned in this Unit. Your goal was to understand how scientists solve problems. You should start thinking about yourself as a student scientist. The things you are learning about how scientists solve problems will help you solve problems in the classroom and outside of school.

Write a brief answer to each question. Use examples from class to justify your answers. Be prepared to discuss your answers in class.

1. *Teamwork*—Scientists and designers often work in teams. Think about your teamwork. Record the ways you helped your team during this Unit. What things made working together difficult? What did you learn about working as a team?
2. *Learning from other groups*—What did you learn from other groups? What did you help other groups learn? What does it take to learn from another group or help another group learn? How can you make *Plan* and *Solution Briefings* work better?
3. *Informed decision making*—What is an informed decision? What kinds of informed decisions did you have to make recently? What do you know now about making informed decisions that you did not know before this Unit? What role do results from investigations play in making informed decisions? Provide an example of using results from an investigation to make a decision during this Unit.
4. *Iteration*—Simply trying again is not enough to get to a better solution or understanding. What else do you need to do to be successful? What happens if your design does not work well enough the second time? What if a procedure you are running does not work well enough the second time?

Reflect

20 min.

Students participate in a discussion that reflects on the main goals of this Unit and answers the Big Question.

△ Guide and Assess

Use the *Reflect* questions to assess students' understanding of the *Basketball-Court Challenge*. Have students write their answers to the *Reflect* questions, and then lead a discussion of students' responses. Use the following points to guide the discussion and assess students' answers.

1. Students' responses should include the contribution of ideas, whether they were used or not, and choices that students made. Students might list difficulties when they have opposing ideas, when they do not understand a group member's idea, or when they misunderstand each other's ideas. Students may mention they learned the usefulness of brainstorming with their group, and ideas that do not work have value and can be used to build on each other's ideas.

2. Make sure students' responses contain the idea of building on each other's ideas and the usefulness of getting advice from the class, particularly in the *Plan* and *Solution Briefings*. Students should also describe their ideas on how to improve these briefings. Remind students of the importance of giving credit. Students may describe how using information from each other's experiments helped to save time so they did not have to investigate everything on their own.

3. Students should define an informed decision as one based on evidence from experiments and science knowledge. In making recent informed decisions, students should describe decisions they made in the challenges to improve the design. Students should mention they used the results of their own experiments and science knowledge in the *Basketball-Court Challenge*.

4. Students should describe how using the iterative process helped improve their designs. They should mention changing one small thing in their design during each iteration as the most effective way of using iteration. If they change many things at once, they will not know which of those things made a difference. Similarly, students should realize if their design does not work on the second try they should not scrap it, but continue modifying it in small ways.

5. Students should know that a criterion is something that must be achieved to satisfy the requirements of the challenge. Students should mention constraints as well. If they do not, bring up constraints, such as the materials, as limitations.

6. Students' answers should say fair tests look at certain factors being tested (under the same conditions.) Procedures must be consistent. Only the manipulated and responding variables change and all other variables must be held constant. They should provide an answer to the investigative question being asked.

5. *Achieving criteria*—What is a criterion? How do you know which criteria are important? What if you cannot achieve all of them? How did you generate criteria? On which challenges were you able to achieve the whole set of criteria? How did you decide which ones to achieve?
6. *Running experiments and controlling variables*—What does it mean to do a fair test? What is hard about doing a fair test? What happens if you do not control important variables? Some variables are more important to control than others. Why? Use examples from class to illustrate. What did you learn about running experiments successfully that you did not know before? Use examples from class to illustrate your answer.
7. *Modeling and Simulation*—Sometimes a process in the world is too small or too large or too complicated to examine. When scientists need to study the process anyway, they often create models and then run simulations on the models. What modeling did you do in this Unit? You used modeling for three different purposes. What purposes did you use modeling for in this Unit? Simulation means running a model. What kinds of simulation did you do? Modeling and simulation are useful only if the model is similar to the real world in important ways. How did you make sure your models were similar enough to the real world for you to be able to learn from them? How did you make sure you were simulating rainfall in a way that was similar enough to the way it happens in the real world?
8. *Using cases to reason*—Scientists and engineers often have to solve problems that others have confronted before. When scientists and engineers work to solve a problem, they may write up their experience as a case study for others to learn from. They will even write up the case study if their solution did not work. This way, others can learn what not to do to solve the problem. You used case studies several times in this Unit. What are the benefits of using case studies to help you solve a problem?

7. Students should understand that when they are investigating something, they cannot determine what causes a change if more than one thing varies. They should only intentionally change one variable, measure how the responding variable changes, and keep all other variables constant if possible. If they cannot keep other variables constant, their change is insignificant to their results. Students should describe examples from the *Build a Boat Challenge* and the *Basketball-Court Challenge*.

8. Students should discuss the ways they built on the ideas and solutions of others, and the benefits. They should describe how building on each other's ideas helps to achieve a better solution and requires students to be attentive to other students as they present their ideas in class and to be aware of the goals. Students should include connections of what they did during any of the challenges of this Unit.

Ask students to answer the *Big Question: How do scientists work together to solve problems?*

◇ Evaluate

Students' responses should include all eight of the items listed: Teamwork; Learning from other groups; Informed decision-making; Iteration; Achieving criteria; Running Experiments; Controlling variables; Using cases to reason.

Teacher Reflection Questions

- What lessons from the *Basketball-Court Challenge* did students apply to interpreting what the *IDEO* design team did?
- How did you assess students' participation in the *Stop and Think* discussion? Is there anything you could do differently next time?
- How did you engage students in watching the video closely and actively, rather than passively?

NOTES

Blackline Masters

Blackline Masters

Digging In Blackline Masters

Name of BLM **Location***

* Number indicates Learning Set.section.sequence within section

Boat Records

Name: ______________________________ **Date:** ______________

Each time you test the design of your foil boat, complete this page. If you change the design of the boat, be sure to fill out a new page.

Draw a simple sketch of your design idea or model.
How many keys did the boat hold? How long did it stay afloat?
What is an advantage to using this design?
What is a disadvantage to using this design?

Solution-Briefing Notes

Name: ______________________________ Date: ______________

Design Iteration: ______________________________

Design or group	How well it works	What I learned and useful ideas		
		Design ideas	Construction ideas	Science ideas

Plans for our next iteration

Lava Flow Data

Name: ______________________________ Date: ______________

Place an X in the box above the number of seconds it took for each group's lava to flow across the plate (starting with Row 1). If more than one group gets the same result, place the X in the next row above the number of seconds.

Lava Flow Bar Graph

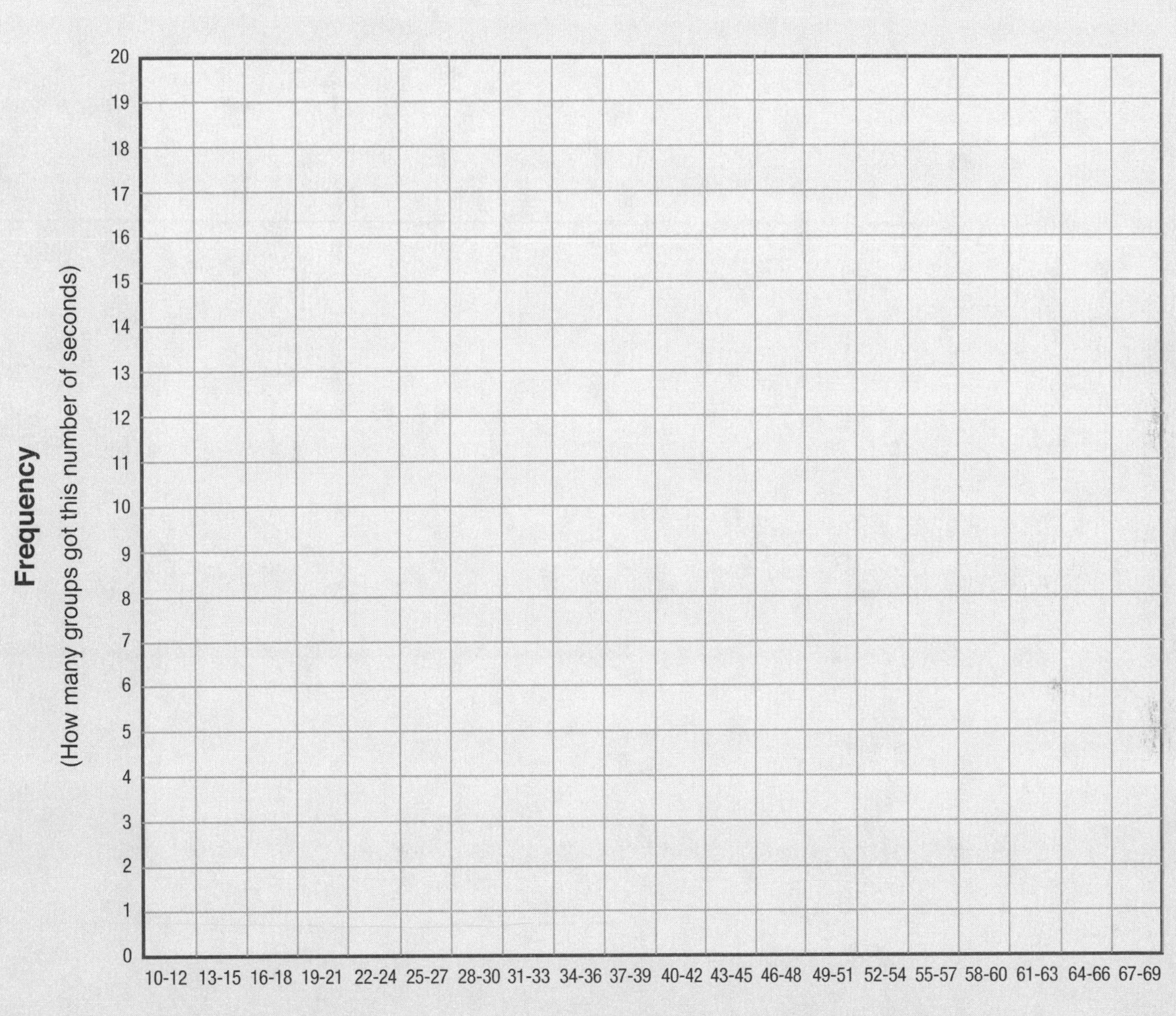

Erosion-Walk Observation

Name: ______________________ **Date:** ______________

Location: ______________ Description: ______________ Cause: ______________	Location: ______________ Description: ______________ Cause: ______________
Location: ______________ Description: ______________ Cause: ______________	Location: ______________ Description: ______________ Cause: ______________

Based on your walk, make a list of the places you think erosion is most likely to occur.

Based on your walk, make a list of causes of erosion.

Erosion Case Study

Name: ______________________________ **Date:** ______________

Case name:	
Case information: time, location	
The situation	
The setup: What was happening before the erosion? What kinds of things was the land being used for?	The problem: What erosion happened, and what problems, if any, did the erosion cause?
The erosion	
Causes: What caused the erosion?	Time: How long did it take for the erosion to happen?
Solutions	
Fixes: How did they try to fix it? Who tried to fix it?	Outcomes: What, if anything, happened as a result?
Conclusions: What can we learn from this case?	
About causes of erosion?	Anything else?

Particle Size and Erosion

Name: ______________________________ Date: ______________

Particle in mixture	Particle size rank (largest = 1 to smallest = 4)	Distance material spread	Patterns of eroded material	Effects on hill
Gravel				
Sand				
Slate				
Soil				

Slope and Erosion

Name: ______________________ Date: ____________

Slope of tray	Distance material spread	Patterns of eroded material	Effects on hill	Other observations
No slope				
Gentle slope				
Steep slope				

Create Your Explanation

3.4.1/3.8.1
3.10.2

Name: ______________________ **Date:** ______________

Use this page to explain the lesson of your recent investigations.

Write a brief summary of the results from your investigation. You will use this summary to help you write your Explanation.

Claim – a statement of what you understand or a conclusion that you have reached from an investigation or a set of investigations.

Evidence – data collected during investigations and trends in that data.

Science knowledge – knowledge about how things work. You may have learned this through reading, talking to an expert, discussion, or other experiences.

Write your Explanation using the ***Claim***, ***Evidence***, and ***Science knowledge***.

Case Summary

Name: ______________________ **Date:** ____________

Case name	
Case description (include problem)	
Case solution (describe and sketch)	
Reason(s) that solution was used	
Outcomes (expected, unexpected, good, problematic)	
Conclusion (What did we find out about erosion management?)	

Erosion Control Model

Name: ______________________ **Date:** ______________

What we are modeling:
Our models' parts (include evidence and science knowledge): Design decision — Reason
Similarities and Differences: How our model is like the real world: How our model is different from the real world:
How we will build our model: Description:
Sketches:

Our Simulation

Name: ______________________ **Date:** ______________

What we are investigating:	
Step-by-step simulation procedure:	
How our procedure is similar to the real world (in materials, design, scale, etc.):	
Description of model setup at start:	Sketch of model at start:
Description of our prediction:	Sketch of our prediction:
How we will vary the procedure to answer our investigation questions:	

Running Our Erosion-Control Model

Name: ______________________ **Date:** __________

Description of setup:	Sketch of setup:

Explain any changes you made to the procedure:

How our procedure is similar to the real world (in design, scale, materials, amount of moisture, etc.):

Description of what happened:	Sketch of what happened:

Summary of measurements and observations:

What we learned and what else we need to investigate:

Conclusions and Recommendations:

Our Design Plan

Name: ______________________ Date: ______________

Design Decisions	Supports	
	Evidence	Science Knowledge

Diagrams

Testing My Design

Name: ______________________ Date: ______________

Each time you build and test a design idea or model, you need to test it in a fair way and record the results of that test. Use this sheet to help record your various design ideas and the result of each design.

Sketch your design. The sketch should help others clearly understand what you are modeling and how it is similar to the real-world situations it models. Draw more than one sketch if you need to.	What is the key idea you are investigating in this simulation?
	How are you running your simulation? How are you making sure you are simulating the real world?
What happened when you ran your simulation? How effective was your design at accomplishing its task? Include measurements and sketches as needed.	What have you learned from this simulation?
	What do you need to investigate to get to a better solution?

Credits

84 Business Park Drive, Armonk, NY 10504
Phone (914) 273-2233 Fax (914) 273-2227
www.its-about-time.com

Publishing Team

President
Tom Laster

Director of Product Development
Barbara Zahm, Ph.D

Managing Editor
Maureen Grassi

Project Development Editor
Ruta Demery

Project Manager
Sarah V. Gruber

Assistant Editors, Student Edition
Gail Foreman
Susan Gibian
Nomi Schwartz

Assistant Editors, Teacher's Planning Guide
Kelly Crowley
Edward Denecke
Heide M. Doss
Jake Gillis
Rhonda Gordon

Creative Director
John Nordland

Production/Studio Manager
Robert Schwalb

Production
Sean Campbell

Illustrator
Dennis Falcon

Technical Art/Photo Research
Sean Campbell
Michael Hortens
Marie Killoran

Equipment Kit Developers
Dana Turner
Joseph DeMarco

Safety and Content Reviewers
Edward Robeck
Barbara Speziale

NOTES

NOTES

NOTES

NOTES

NOTES

NOTES

NOTES

NOTES